Super-Resolution Microscopy for Material Science

Optical microscopy is one of the most frequently used tools in chemistry and the life sciences. However, its limited resolution hampers the use of optical imaging to many other relevant problems in different disciplines. Super-Resolution Microscopy (SRM) is a new technique that allows the resolution of objects down to a few billionth of meters (nanometers), ten times better than classical microscopes, opening up opportunities to use this tool in new fields.

This book describes the theory, principles, and practice of super-resolution microscopy in the field of materials science and nanotechnology. There is a growing interest in the applications of SRM beyond biology as new synthetic materials, such as nanoscale sensors and catalysts, nanostructured materials, functional polymers, and nanoparticles, have nanoscopic features that are challenging to visualize with traditional imaging methods.

SRM has the potential to be used to image and understand these cutting-edge man-made objects and guide the design of materials for novel applications.

This book is an ideal guide for researchers in the fields of microscopy and materials science and chemistry as well as graduate students studying physics, materials science, biomedical engineering, and chemistry.

Key Features:

- Contains practical guidance on Super-Resolution Microscopy (SRM), an exciting and growing tool that was awarded the Nobel Prize for chemistry in 2014.

- Provides a new perspective targeting materials science, unlike existing books which target readers in chemistry, life science, and biology.

- Targets students in its core chapters, while offering more advanced material for professionals and researchers in later chapters.

Super-Resolution Microscopy for Material Science

Edited by
Lorenzo Albertazzi and Peter Zijlstra

CRC Press
Taylor & Francis Group
Boca Raton London New York

CRC Press is an imprint of the
Taylor & Francis Group, an **informa** business

Designed cover image: Adapted from a microscopy image acquired by
Manos Archontakis

First edition published 2024
by CRC Press
6000 Broken Sound Parkway NW, Suite 300, Boca Raton, FL 33487-2742

and by CRC Press
4 Park Square, Milton Park, Abingdon, Oxon, OX14 4RN

CRC Press is an imprint of Taylor & Francis Group, LLC

ISBN: 978-1-032-10367-9 (hbk)
ISBN: 978-1-032-11608-2 (pbk)
ISBN: 978-1-003-22068-8 (ebk)

DOI: 10.1201/9781003220688

Typeset in Minion
by SPi Technologies India Pvt Ltd (Straive)

Contents

Editors

Lorenzo Albertazzi is Associate Professor at Eindhoven University, The Netherlands, within the department of Biomedical Engineering. He earned an MSc in Chemistry (2007) and a PhD in Biophysics (2011) from Scuola Normale Superiore (Pisa, Italy). He then joined Eindhoven University of Technology as a postdoctoral researcher and, in 2014, became a NWO/VENI fellow. In 2015, he moved to Barcelona (Spain) to the Institute of Bioengineering of Catalonia (IBEC) to start the 'Nanoscopy for Nanomedicine' group. Since 2018 he has been an associate professor at TU/e, leading the research group Nanoscopy for Nanomedicine.

Peter Zijlstra is an Associate Professor at Eindhoven University, The Netherlands, in the research group Molecular Biosensing in the department of Applied Physics. He studied Applied Physics at the University of Twente (Enschede, The Netherlands), where he earned his MSc degree in 2005. In 2009, he earned his PhD from Swinburne University of Technology (Melbourne, Australia), where he studied the photothermal properties of single plasmonic nanoparticles, with applications in optical data storage. After a postdoctoral fellowship in the lab of Prof. Michel Orrit at Leiden University (The Netherlands) he moved to Eindhoven University of Technology (TU/e, The Netherlands). He is a core member of the Institute for Complex Molecular Systems at TU/e, where groups from different disciplines (chemistry, physics, biomedical engineering, and mathematics) collaborate on multidisciplinary research topics.

Contributors

Guillermo P. Acuna
Department of Physics, University of Fribourg
Fribourg, Switzerland

Sarit S. Agasti
New Chemistry Unit, Jawaharlal Nehru Center for Advanced Scientific Research (JNCASR)
Bangalore, India
and
Chemistry & Physics of Materials Unit, Jawaharlal Nehru Center for Advanced Scientific Research (JNCASR)
Bangalore, India
and
School of Advanced Materials (SAMat)
Jawaharlal Nehru Center for Advanced Scientific Research (JNCASR)
Bangalore, India

Lorenzo Albertazzi
TU/e Department of Biomedical Engineering
Eindhoven University of Technology
Eindhoven, The Netherlands

Andrea Baldi
Department of Physics and Astronomy
Vrije Universiteit Amsterdam
Amsterdam, The Netherlands

Hans-Dieter Barth
Goethe-University Frankfurt
Frankfurt, Germany

Marina S. Dietz
Goethe-University Frankfurt
Frankfurt, Germany

Florencia Edorna
Centro de Investigaciones en Bionanociencias (CIBION)
Consejo Nacional de Investigaciones Científicas y Técnicas (CONICET)
Ciudad Autónoma de Buenos Aires, Argentina
and
Departamento de Física, Facultad de Ciencias Exactas y Naturales
Universidad de Buenos Aires
Ciudad Autónoma de Buenos Aires, Argentina

Deepika Gupta
New Chemistry Unit, Jawaharlal
    Nehru Center for Advanced
    Scientific Research (JNCASR)
Bangalore, India

Ruben F. Hamans
Department of Physics and
    Astronomy
Vrije Universiteit Amsterdam
Amsterdam, The Netherlands

Mike Heilemann
Goethe-University Frankfurt
Frankfurt, Germany

Teun A.P.M. Huijben
Department of Health Technology
Technical University of Denmark
    (DTU)
Kongens Lyngby, Denmark

Simanta Kalita
Chemistry & Physics of Materials
    Unit
Jawaharlal Nehru Center for
    Advanced Scientific Research
    (JNCASR)
Bangalore, India

Lucía F. Lopez
Centro de Investigaciones en
    Bionanociencias (CIBION)
Consejo Nacional de
    Investigaciones Científicas y
    Técnicas (CONICET)
Ciudad Autónoma de Buenos
    Aires, Argentina

Rodolphe Marie
Department of Health Technology
Technical University of Denmark
    (DTU)
Kongens Lyngby, Denmark

Kim I. Mortensen
Department of Health Technology
Technical University of Denmark
    (DTU)
Kongens Lyngby, Denmark

Sílvia Pujals
Department of Biological
    Chemistry
Institute for Advanced Chemistry
    of Catalonia (IQAC-CSIC)
Barcelona, Spain

Frank Scheffold
Department of Physics
University of Fribourg
Fribourg Switzerland

Fernando D. Stefani
Centro de Investigaciones en
    Bionanociencias (CIBION)
Consejo Nacional de
    Investigaciones Científicas y
    Técnicas (CONICET)
Ciudad Autónoma de Buenos
    Aires, Argentina
and
Departamento de Física, Facultad
    de Ciencias Exactas y Naturales
Universidad de Buenos Aires
Ciudad Autónoma de Buenos
    Aires, Argentina

Ilja Voets
Laboratory of Self-Organizing Soft
    Matter
Department of Chemical
    Engineering and Chemistry, and
    Institute for Complex Molecular
    Systems
Eindhoven University of
    Technology
Eindhoven, The Netherlands

Piotr Zdańkowski
Centro de Investigaciones en
    Bionanociencias (CIBION)
Consejo Nacional de
    Investigaciones Científicas y
    Técnicas (CONICET)
Ciudad Autónoma de Buenos
    Aires, Argentina

and

Warsaw University of Technology
Institute of Micromechanics and
    Photonics
Warsaw, Poland

Peter Zijlstra
TU/e department of Applied
    Physics
Eindhoven University of
    Technology
Eindhoven, The Netherlands

Introduction to Super-Resolution Microscopy and Its Importance for Materials Science

Lorenzo Albertazzi and Peter Zijlstra

Eindhoven University of Technology, Eindhoven, The Netherlands

1.1 INTRODUCTION

Microscopy and other characterization techniques play a pivotal role in materials science. After the synthesis or formulation of a new material, it is imperative to understand the properties and key features of this new product to verify if the intended material has been created and to obtain quantitative information about its physicochemical properties. Moreover, imaging techniques are crucial to understand the functioning of materials in situ (i.e., in the context of its application). This is growing in importance because complex and dynamic materials can now be synthesized that pose a real challenge toward their understanding. This has reached the point at which understanding and characterization are the main hurdles that stop the development of new materials rather than our ability to synthesize them. Therefore, the quote from the 2002 Nobel Prize Sydney Brenner, "progress in science depends on new techniques, new discoveries and new ideas, probably in that order" cannot be more timely in our field.

DOI: 10.1201/9781003220688-1

The arsenal of techniques available to chemists increased at an incredible rate in the last 10 years, with new methods available that have the potential to unveil details and information inaccessible before. It is also interesting to note that methods are crossing over between disciplines (for example, methods initially designed to study biological systems are now translated toward chemistry and materials science applications).

In this book we will discuss how the advent of fluorescence super-resolution microscopy (SRM), or nanoscopy, revolutionized our approach to studying materials. This book will describe the principles and applications of this family of techniques and may also serve as a guide to make an informed choice among the large variety of super-resolution methods available and the jungle of imaging acronyms that the reader will find in the literature. The latter is a crucial point. Any material, any scientific question is often best addressed with a different set of techniques, while there are no one-size-fits-all solutions. There is no "best technique" but only the best technique for a specific situation. Being able to make the right choice can make the difference between a failed experiment and a successful material characterization.

1.2 FLUORESCENCE AND FLUORESCENCE MICROSCOPY

Light microscopy (LM) has been widely used in the past in the fields of materials thanks to its minimal invasiveness, multicolor ability, and specific labeling. However, it has be noted that LM did not occupy role like in life sciences, where the optical microscopy is probably the most used instrument. The reason has to be found in the limited resolution imposed by the diffraction limit and the wavelength of visible light that does not reach the nanometer level necessary to study the molecular nature of synthetic materials. For this reason, techniques like atomic force microscopy (AFM) or electron microscopy (EM) are often preferred. Super-resolution microscopy has the potential to change this paradigm, allowing the ability to reach resolutions down to 1 nm while keeping some of the original advantages of fluorescence microscopy. In this section we will provide the basic principles of classical fluorescence microscopy, the stepping-stone toward nanoscopy.

1.2.1 Fluorescence and Key Fluorophore Features

Fluorescence is the emission of light from the excited state of a substance that has previously absorbed light [1]. When an atom or a molecule absorbs light, this energy brings it from the lower energy state, also known as ground state, to a higher energy level, also called excited state. This is best visualized in a Jablonski diagram (see Figure 1.1a), a representation of a selection of the energy levels of a molecule against their energy. Typically, in the Jablonski diagram, only the electronic states (S) and the corresponding vibrational states (0,1,2...) are represented. Notably, fluorescence and absorption have different wavelengths, as some energy is dissipated between vibrational states. As a result, the emission is always redshifted compared to the absorption, and the difference between the two is generally called Stokes shift in honor of George Gabriel Stokes, the first to observe and describe the phenomenon of fluorescence. This has important ramifications for the fluorescence microscope as it implies that the excitation and the emission can be separated using wavelength filters in the optical path. Moreover, this is also crucial for techniques that make use of stimulated emission like STED (see Chapter 3) and require dyes with a sizeable Stokes shift. The presence of multiple vibrational states in molecules both in the ground and the excited states determines that room-temperature fluorescence spectra are not made of lines, like in the case of atoms, but rather broad peaks (see Figure 1.1b).

Of course, fluorescence is not the only way to release energy from an excited state. Many non-radiative (nr) processes can occur, such as internal conversion (the release of energy via heat) or energy transfer to another molecule. All these phenomena have a characteristic rate k and

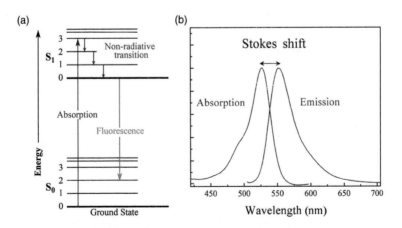

FIGURE 1.1 (a) Jablonski diagram of energy states of a molecule and related transitions, including fluorescence. (b) Typical absorption and fluorescence emission spectra highlighting the Stokes shift.

the probability of fluorescence to happen (i.e., the fluorescence quantum yield) can be defined as;

$$\phi_F = \frac{\#\,\text{photons emitted}}{\#\,\text{photons absorbed}} = \frac{k_r}{k_r + \Sigma k_{nr}},$$

where k_r is the radiative decay rate (green arrow in Figure 1.1a) and k_{nr} the total non-radiative decay rate due to all non-radiative processes. ϕ_F is an important feature of fluorescent molecules and together with the absorption cross section σ_{abs} (units m^2) determines the brightness B (in units photons/s) at moderate excitation power densities:

$$B \propto \sigma_{abs} \Phi_F I_{exc},$$

where I_{exc} is excitation fluence in units of photons/s/m^2. Another important parameter characteristic of a fluorophore is the photostability. When a dye is continuously irradiated, it shuttles between the ground and the excited state, but when it occupies the latter some other photochemical process can occur. In particular it is important to take into account that excited states are more redox active; therefore, they are most subject to photo-oxidation. Oxidation reactions result in the permanent chemical alteration of the structure of the chromophore that destroys its ability to fluoresce, and they are indicated as photobleaching. The typical photophysical path toward this photobleaching generally involves a transition to a long-lived triplet state via intersystem crossing and then a chemical reaction with triplet oxygen present in solution. The photostability, i.e., the ability of a fluorophore to resist to photobleaching, is a key intrinsic feature of a chemical structure and is very important for optical microscopy: more photostable dyes will provide signal for longer times and can withstand higher irradiation powers without losing signal. This is particularly relevant for the super-resolution field, as most SRM methods employ high excitation power and require very photostable dyes. While photobleaching is an irreversible transition toward a dark state, it is also possible to have reversible phenomena that are generally indicated as blinking. The type of states involved in blinking and their corresponding lifetimes can vary enormously, from microseconds to minutes. Blinking typically occurs when the dye transits to a longer-lived state (for example the triplet state), during which it cannot be excited until it has emitted a phosphorescence photon to decay back to the ground state. Once back in the ground state, the dye can cycle between the single ground and excited states to yield

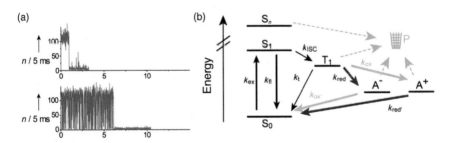

FIGURE 1.2 (a) Example of molecule traces in time for individual dyes showing a dye that has no blinking and undergoes fast photobleaching (top) and a dye that has significant blinking events and a more extended lifetime before photobleaching. (b) Jablonski diagram showing the complex photophysics of a dye, including several dark states resulting in redox reactions in the excited state (A^- and A^+). (Reprinted with permission from [2].)

fluorescence, with dark periods in between while the dye resides in the longer-lived triplet state.

While blinking is generally detrimental for imaging (e.g., reducing the fluorescent signal), it can also be exploited, and it is at the basis of several single molecule localization techniques such as Stochastic Optical Reconstruction Microscopy (STORM). Note that both bleaching and blinking are very sensitive to environmental conditions and therefore cannot be fully avoided but can be tuned by an optimal choice of the buffer (e.g., pH, oxygen concentration, presence of reductive species).

Figure 1.2a shows the signal of individual molecules detected in time that undergoes multiple reversible blinking events before their irreversible photobleaching. Different traces correspond to different buffer conditions, showing dramatic changes in photobleaching or blinking rates. This blinking process due to triplet state dynamics can be described using an extended Jablonski diagram as shown in Figure 1.2b, where the triplet state T1, reduced and oxidized states A^- and A^+, and photobleaching pathways B are now explicitly included.

1.2.2 Fluorescent Labels

Several chemically different families of fluorophores are available and currently used for fluorescence microscopy. Similarly to microscopy techniques there is nothing like the "best fluorophores," but every type has advantages and disadvantages and their knowledge is important in order to make an informed choice. Due to their relevance for super-resolution microscopy, we will discuss organic dyes, fluorescent proteins, and quantum dots.

Organic dyes are small (500–1500 Da, around 1 nm in size) organic molecules that exhibit fluorescence. Since the first synthesis of fluorescein by Adolf van Bayer in 1871 [3], an extremely wide variety of chemical structures have been produced spanning the whole visible spectra, changing in brightness and other photophysical properties.

Figure 1.3 shows examples of the chemical structures of some of the most used fluorophores; despite their chemical diversity, it is clear that

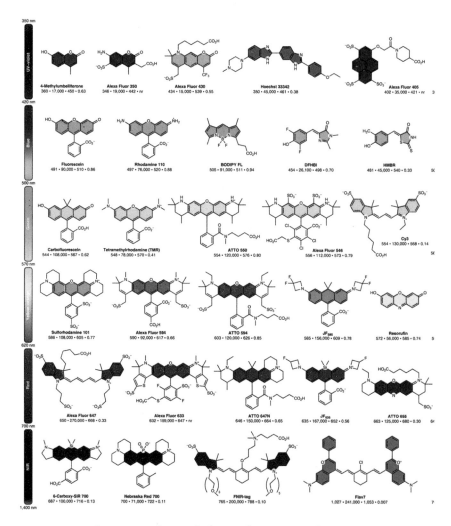

FIGURE 1.3 Overview of typical chemical structures for organic fluorescent molecules across the visible range. Excitation wavelength, extinction coefficient, emission wavelength, and quantum yield are reported. (Reprinted with permission from [4].)

there are common structural features. Firstly, all the chromophores exhibit a conjugated system of double bonds, necessary to absorb the light in the visible range. Secondly, all the dyes have multiple fused rings that ensure rigidity, favoring the deactivation of the excited state via light emission rather than via the molecular vibration associated with heat release. Finally, multiple charges are present to improve solubility and to promote charge-transfer energy transitions. Among the advantages of organic dyes, their small size is paramount; generally, they are significantly smaller than the molecule that they label, minimizing perturbation of the system under examination. Moreover, dyes can be finely tuned in their properties by chemical modifications resulting in fluorophores with optimal photophysical properties (e.g., high brightness and photostability) and allowing them to be tuned when very specific features are required to perform a super-resolution method (e.g., tuned blinking for STORM).

Nowadays, when opening the catalog of a dye manufacturer, a microscopist can easily find a colorful cornucopia of organic fluorophores and choose the most suitable for the desired use. Of course, all these desirable properties come at a price. The rigid aromatic structure of organic dyes makes them rather sticky and hydrophobic. This can induce perturbation of the tagged molecule (for example, making it less soluble or adding unspecific undesired interactions with the surroundings). This is particularly important for self-assembled materials, where the fine balance of hydrophobic interactions is crucial for their final structure. Therefore, the choice of the dye is crucial, and often several attempts to find a suitable one have to be made. Moreover, it is crucial to check if the material key properties are not altered by an independent non-fluorescence-based method (e.g., checking material morphology by AFM [5]).

Another relevant family of dyes is represented by fluorescent proteins (FPs). The discovery of this chromophore began around 1960 when Osamu Shimomura studied a Pacific Ocean jellyfish that exhibit a surprising spontaneous green: the *Aequorea Victoria* (See Figure 1.4a). It took decades before the gene responsible was identified and the protein isolated and named Green Fluorescent Protein (GFP). Shimomura determined the structure of the chromophore: a rigid organic molecule, resembling the structure of organic dyes, embedded in the tridimensional beta-barrel structure of the protein (Figure 1.4b). A key advantage of FPs is that they are fully genetically encoded. Indeed the sequence of DNA coding for FPs can be delivered to other cells or entire organisms completely diverse from the jellyfish, and it will express the fluorescent

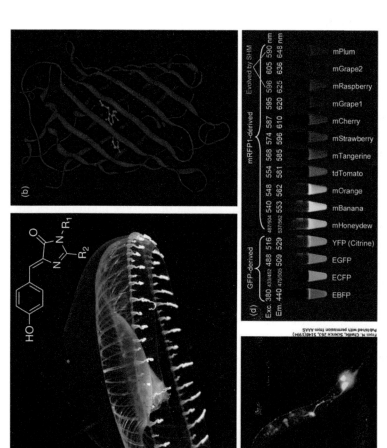

FIGURE 1.4 (a) Picture of the luminescent jellyfish Aequorea Victoria with the GFP chromophore depicted in white. (b) 3D structure of the GFP protein showing the beta-barrel in blue and the chromophore in green. (c) Expression of the GFP in a C. Elegans. (d) palette of GFP mutants of different colors. (Adapted with permission from [8, 9].)

protein without the need of any cofactor, as brilliantly demonstrated by Martin Chalfie in the worm *C. Elegans* [6] (Figure 1.4c). Moreover, the FPs can be expressed together with another protein into a fluorescent chimera, allowing the easy labeling of proteins in vitro, in cells, and *in vivo*. Another interesting advantage of FPs is that their photophysical properties can be tuned by mutating amino acids in the protein sequence. This paved the way toward the screening of chromophores that covers the whole spectra from blue to red and with enhanced brightness (Figure 1.4d). For their ease of protein labeling and biocompatibility, FPs are often the preferred dyes in the cell biology, chemical biology, or biomaterial fields [7].

However, FPs are not deprived of disadvantages. The bulky protein structure (4 × 2 nm) may cause steric hindrance and change the behavior of the labeled object. Moreover, they still exhibit suboptimal brightness, blinking, and photostability compared to organic dyes.

The last type of chromophore with relevance for super-resolution microscopy are quantum dots (QDs). QDs are nanometer-sized semiconductor particles with optical properties that differ from the bulk materials due to quantum mechanical size-effects (Figure 1.5a). QDs are generally excited with UV light, bringing one electron from the valency band to the conductance band. The resulting electron-hole pair emits fluorescence at a wavelength that depends on the size and material of the QD. Interestingly the emission spectrum of QDs depends on their size and shape as they influence the energy gap between the valency and conductance band. For example, 6 nm spherical QDs emit in the red region while 2 nm QDs in the

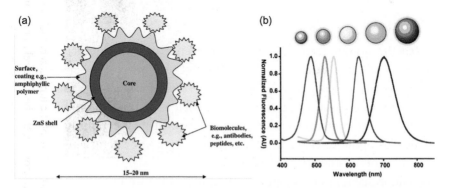

FIGURE 1.5 (a) Schematic representation of a quantum dot for bioimaging highlighting the different layers of the material and their contribution to the total size. (b) Spectra of fluorescent quantum dots of different size showing the effect of size on emission color. ([a] Reprinted with permission from [10]; [b] Reprinted with permission from [11].)

TABLE 1.1   Performances of the Three Family of Fluorophores and SRM Methods Mostly Associated with

	Brightness	Photostability	Ease of Labeling	Size	Tunability	Biocompatibility	SRM Method Mostly Used for
Organic dyes	+	++	++	+++	+++	+	All
Fluorescent proteins	-	-	+++	-	++	+++	PALM, RESOLFT
Quantum dots	+++	+++	-	-	+	-	SOFI, STED

blue. Therefore, the synthesis of QD of tuned size and morphology results in a palette of labels for bioimaging (Figure 1.5b). Another interesting feature of QDs is their unique photophysics. Quantum dots are extremely photostable and virtually never undergo photobleaching. However, they generally exhibit blinking that on one side limits their use when continuous emission in time is needed while on the other make them suitable for techniques where blinking is required (e.g., STORM or Super-Resolution Optical Fluctuation Imaging, SOFI). Moreover, they are extremely bright compared to molecular fluorophores due to their high absorption cross section. Their biggest disadvantages are their bulky nature (5–20 nm), their difficult labeling procedure, and some semiconductor materials exhibit toxicity for the field of biomaterials.

Overall advantages and disadvantages of the different fluorophore types are listed in Table 1.1 that may serve as a guide to choose the best dye for the given research question.

1.2.3 Fluorescence Microscope Layout

Given the unique features of the fluorescence process described in the previous sections, different types of fluorescence microscopes have been designed and are currently used for super-resolution imaging. They differ in the geometry of the optical system as well as in the choice of components such as excitation source and detectors. In this section the general principles of fluorescence microscopy and some key microscopy designs used in super-resolution microscopy will be discussed. The general design for a fluorescence microscope is highlighted in Figure 1.6a.

A light source (e.g., a white lamp, a led, or a laser) is used to excite a fluorescently labeled sample. The light from the source passes through an excitation filter that selects the desired excitation wavelength for the fluorophore of interest. The beam then passes through the "heart" of the

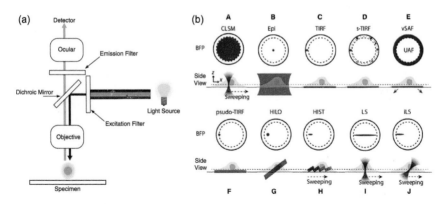

FIGURE 1.6　(a) schematic depiction of the general architecture of a fluorescence microscope. (b) Overview of the illumination schemes for fluorescence microscopy showing the pattern of the beam in the back focal plane (BFP) and the resulting illumination of the sample. (Reprinted with permission from [12].)

microscope, the dichroic mirror. This optical component behaves like a mirror for specific wavelengths while being nearly transparent for others. Knowing that due to the Stokes shift the excitation and emission of fluorescence occur at different wavelengths, the dichroic mirror can effectively separate the two, avoiding contamination of the image with excitation photons. Typically, the excitation light is reflected by the mirror and the beam is directed to the sample with an objective lens, where it excites the fluorescence. The fluorescent signal is then typically collected by the same objective lens and directed back to the dichroic mirror. Here the dichroic mirror lets the fluorescence photons through while it blocks all the other wavelengths, including the excitation photons. A further emission filter ensures that only fluorescence is detected and not scattering or other undesired signals at the wavelength of the source. Finally, the light reaches the detector, typically a camera or a single point detector such as a photomultiplier tube, where the image is built.

A variety of fluorescence microscopes, suitable for different super resolution techniques, can be built by changing the pattern of illumination of the excitation light, i.e., the shape and collimation of the beam reaching the back focal plane (BFP) of the objective lens as shown in Figure 1.6b. All of these excitation modalities are used in SRM but three are the most common: epi-, confocal-, and TIRF-illumination. (See Figure 1.7.)

In epifluorescence (Figure 1.7a), the excitation beam is focused on the back focal plane of the objective lens, resulting in a collimated beam

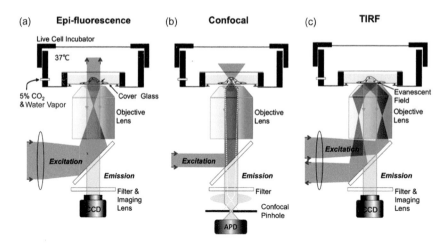

FIGURE 1.7 Schematic depiction of the general architecture of a fluorescence microscope using (a) epifluorescence, (b) confocal, and (c) total-internal reflection fluorescence (TIRF) modalities. (Adapted with permission from [13].)

emanating from the lens that excites the whole sample homogenously with light (Figure 1.6b(B)). The fluorescence from the whole field of view (FOV) is collected by the objective and the spatial image is resolved on a CCD or CMOS camera. This is a simple and robust optical design that allows for fast imaging because the whole FOV is measured at once. However, it also results in limited signal-to-background ratios because out-of-focus fluorescence is also measured. This illumination is often used for SIM, fluctuation-based SRM methods such as SOFI and Super-Resolution Radial Fluctuations (SRRF), and less often for SMLM methods such as STORM (these methods are discussed Chapter 2).

With the opposite approach to epifluorescence, confocal microscopes (Figure 1.7b) illuminate the back focal plane of the objective lens with a collimated beam, resulting in a strongly focused illumination spot on the sample (Figure 1.6b(A)). To build an image this focal point is scanned across the sample and the fluorescence collected through a detector one point at a time. To reject all light not coming from the focal plane (indicated by the dashed lines in Figure 1.7b) a pinhole is placed in front of the detector that acts as a micrometric spherical aperture that blocks out-of-focus light. This ensures that any signal not originating from the focal plane is largely blocked, resulting in a good signal-to-background ratio. Importantly, this geometry also allows for 3D imaging because such optical sectioning ensures that only signal from one z-plane passes the pinhole. The use of point scanning comes at the cost of temporal resolution as the scanning slows down image acquisition, while potential photodamage may occur in sensitive samples due to the strongly focused excitation beam. STED super-resolution microscopes are based on a confocal design and can also be used in combination with other fluctuation-based methods such as SOFI and SRRF. STED microscopy is described in detail in Chapter 3.

To obtain high signal-to-noise without losing temporal resolution, total internal reflection fluorescence (TIRF, Figure 1.7c) can be used. In TIRF the excitation light is focused on the back focal plane of the objective, but compared to epifluorescence microscopy, it is displaced sideways to the edge of the objective lens (Figure 1.6b(C)). As such, the light beam illuminates the sample under an angle, having the light emanating from only one side of the objective. If the illumination angle exceeds the critical angle of the glass-sample interface, all the excitation light is reflected at the interface and only an evanescent wave penetrates into the sample. As the evanescent wave decays exponentially in the z-direction, only the first 100–200 nm of the samples are illuminated, resulting in extremely high signal-to-background ratio as the rest of the sample does not receive any light. Moreover, the temporal resolution is also excellent as all the sample is illuminated at once. For these reasons TIRF is the method of choice for techniques where sensitivity is crucial, like in the case of SMLM that is based on the detection of single molecules. The obvious price paid is that only a small part of the sample close to the coverslip is measured and thick samples are not accessible. To improve this, methods such as pseudo-TIRF and HiLo illuminations are used. These illuminate with an angle slightly below the critical angle, resulting in deeper penetration of the excitation light at the cost of higher background levels. Tuning the angle allows finding the optimal compromise between signal-to-noise and depth of imaging, depending on the desired application.

While the decision-making process for the choice of the microscope type is more established in biology, where samples are more standard, this is an open question in materials science. Materials vary more wildly in size (from atomic level to bulk), optical transparency (from transparent to opaque or strongly scattering) and refractive index, making the choice more challenging. While nanomaterials (e.g., nanoparticles) are ideal for TIRF imaging, complex 3-dimensional hierarchical materials need an accurate choice of the illumination method and often require more than one technique in order to obtain a full picture of the molecular architecture.

1.3 BREAKING THE DIFFRACTION LIMIT: SUPER-RESOLUTION MICROSCOPY

1.3.1 The Diffraction Limit

While fluorescence microscopy has many advantages for material characterization, its use was often limited in material science and chemistry due to the limited spatial resolution. Modern optical instruments are not limited anymore by alignment or imperfection of the optics but they are still suffering from the physical limit of diffraction. Ernst Abbe already found in 1873 [14] that light with wavelength λ, traveling in a medium with refractive index n focused to a spot through solid angle θ will result in a minimum resolvable distance of:

$$d = \frac{\lambda}{2n\sin\theta} = \frac{\lambda}{2\mathrm{NA}}$$

Typically, the numerical aperture (NA) of modern objectives can reach 1.4. Taking into account the visible light wavelengths (400–800 nm), this results in typical resolution in the range of a few hundred nanometers. This is related to the wave nature of light that results in a focal spot size that is dictated by diffraction. The wavefronts emitted by a point-like object (e.g., a dye molecule) are collected by the objective lens and focused on the camera. On the camera these wavefronts interfere, resulting in a blurred image of the point-like object. Therefore, a point source gives a signal constituted of a central diffraction spot surrounded by multiple diffraction rings, also referred to as the point spread function (PSF). The size of the central spot is expressed in the Abbe formula above as d and constitutes the limit to the spatial resolution of optical microscopes. When trying to resolve two adjacent objects, both of them will result in an Airy pattern on the camera, and if the two objects are closer than d, the two patterns will overlap and the objects cannot be distinguished. In a real sample, the measured image is a convolution of the actual object and the PSF of the optical system (Figure 1.8).

Another approach to assess the resolution of a microscope is to reason in terms of spatial frequencies that are transmitted by the optics in the light path as shown in Figure 1.9. An object can be represented in Fourier space as the sum of multiple periodic functions with different frequencies, where higher frequencies represent the fine details of the object and the lower frequencies, the coarse features. Performing a Fourier transformation of the PSF results in the optical transfer function (OTF), a description

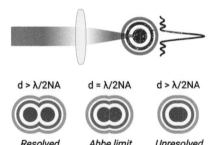

d > λ/2NA   d = λ/2NA   d > λ/2NA

Resolved   *Abbe limit*   *Unresolved*

FIGURE 1.8   Sketch of the definition of the Abbe limit of resolution.

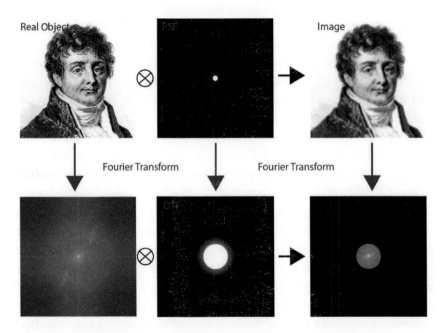

FIGURE 1.9   Schematic showing the relation between image resolution and the point spread function in Fourier space.

of how efficiently different frequencies are transferred to the final camera image. The transfer function approaching zero indicates that these frequencies are no longer efficiently transferred and therefore not detected. Finally, only a finite bandwidth of spatial frequencies is transferred to the camera, resulting in the finite (diffraction limited) resolution. This frequency-space analogy is a more formal and robust definition of resolution and will help the explanation of the principle of some SRM methods like SIM in Chapter 4.

While the diffraction limit cannot be overcome with better lenses or alignment, some clever approaches can be found to bypass the Abbe limit and reach what we now call super-resolution imaging.

1.3.2 Breaking the Diffraction Limit: An Historical Perspective on Super-Resolution Microscopy

The Abbe criteria was so well-formulated that it took a century for scientists to even think to challenge it. First attempts to bypass the diffraction limit dates to the 1990s when the first attempts to improve axial z-resolution by means of multiple objectives and interference were proposed (e.g., 4pi microscopy [15]). Slightly later, the theoretical foundation of super-resolution methods able to improve the lateral xy resolution was reported. These reports are the stepping-stone toward the development of the three most common families of super-resolution methods: stimulated emission depletion (STED) [16, 17], structured illumination microscopy (SIM) [18] and single-molecule localization microscopy (SMLM) [19–21]. However, it took one decade of technological development in laser sources, optics, detectors, as well as new fluorescent markers to turn these theories into real microscopy techniques [22].

In the 2000s, there has been an explosion of practical implementations of super-resolution methods, as well as the first SRM-guided discoveries in the field of cell biology. The field grew enormously in the last year, with scientists on one side improving the existing techniques in spatial and time resolution, usability, and applicability and on the other side using these cutting-edge techniques to answer open questions in cell and molecular biology. While biology has been the driving force behind many super-resolution developments, in the last decade SRM demonstrated its potential beyond proteins and cells. Fields like materials science, supramolecular chemistry, nanotechnology, and nanomedicine are starting to appreciate the advantages of super-resolution microscopy, and these techniques are becoming standard tools for synthetic structures characterization. SRM, or nanoscopy, is still a growing field with new methods being developed and applied. For example, MINFLUX is pushing the boundaries of both time and spatial resolution, reaching the atomic level [23], and the combination with expansion microscopy holds a great potential to further enhance the resolution without using complex and expensive instrumentations [24].

The microscopy market followed up closely with the technological developments in nanoscopy. Since early 2000, the first commercial super-resolution microscopes are available, but it is only from the 2010s that the market grew. Many options are now available from different manufacturers. The commercial availability of SRM had a huge impact on the spread of these methods, turning nanoscopy from a niche technique to a standard must-have for microscopy facilities. Figure 1.10 shows a timeline of these super-resolution milestones.

An interesting development of the field is also the open microscopy approach. In this philosophy, scientists building novel super-resolution setups share all the technical details and designs of their microscopes, together with a detailed guide on how to build them. This increases the accessibility of super-resolution microscopy by reducing microscope cost by 2–5 fold, allows new developers of rapid prototyping, and in some cases helps transparency and reproducibility of super-resolution experiments. Several open super-resolution projects are available, including the miCube [25], k2 TIRF [26], liteTIRF [27], NanoPro [28], and cellSTORM [29]. To further standardize and improve reproducibility, several initiatives such as the QUAREP-LiMi have proposed guidelines for good practice optical microcopy, from the microscope design to the sample preparation and acquisition [30].

1.3.3 Families of Super-Resolution Microscopy: A Guide through the Jungle of Acronyms

As discussed in the previous section, the SRM field has grown exponentially since its introduction, and many techniques and methods are now available. This, together with the love of biophysicists for acronyms, results in a complicated landscapes of technique names: STED, RESOLFT, 4PI, (d)STORM, PALM, PAINT, IRIS, SIM, SRRF, SOFI, MINFLUX, MINSTED, RASTMIN, ROSE, and many others. To guide the beginner in this confusing environment, we decided to group all the methods into four sections: i) STED-like, ii) SMLM-like, iii) SIM-like, and iv) other techniques that do not fall into the previous categories. The next sections will discuss these methods individually, highlighting the basic principles, the technical implementation, and some of their applications. In this section we will provide a general overview, discussing the classification of the super-resolution methods as well as comparing their performances. Figure 1.11 schematically depicts this classification and offers a table of the key performances of the different methods.

The first family of techniques presented in Figure 1.11 are based on structured illumination (SIM). Here, the sample is illuminated with a non-homogenous patterned light and the fluorescence resulting from the

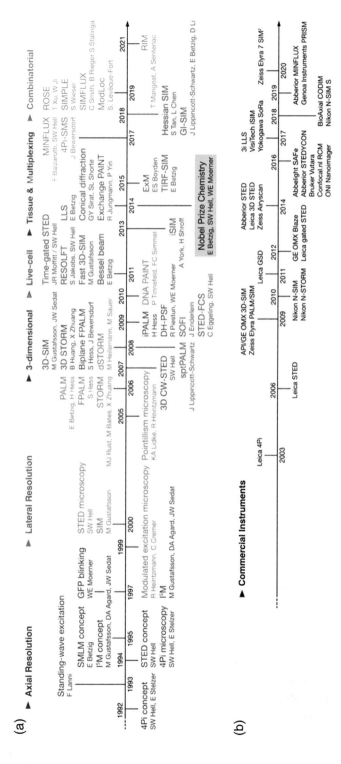

FIGURE 1.10 Historical timeline of the development of super-resolution techniques and their commercially available counterparts. (Reprinted with permission from [22].)

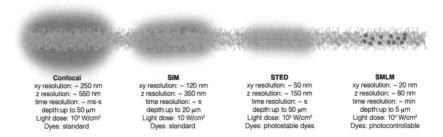

Confocal	SIM	STED	SMLM
xy resolution: ~ 250 nm	xy resolution: ~ 120 nm	xy resolution: ~ 50 nm	xy resolution: ~ 20 nm
z resolution: ~ 550 nm	z resolution: ~ 350 nm	z resolution: ~ 150 nm	z resolution: ~ 80 nm
time resolution: ~ ms-s	time resolution: ~ s	time resolution: ~ s	time resolution: ~ min
depth:up to 50 μm	depth:up to 20 μm	depth:up to 50 μm	depth:up to 5 μm
Light dose: 10^3 W/cm²	Light dose: 10 W/cm²	Light dose: 10^6 W/cm²	Light dose: 10^4 W/cm²
Dyes: standard	Dyes: standard	Dyes: photostable dyes	Dyes: photocontrollable

FIGURE 1.11 Overview of the performances of the different SRM methods. (Adapted with permission from [31, 32].)

interference of the excitation with the labeled sample measured. Upon illumination of the same area with multiple patterns, a mathematical deconvolution is applied to retrieve an image with enhanced resolution. This is a relatively simple method, based on minimal modification of widefield or TIRF illuminations, that is compatible with standard fluorescent markers and uses a minimal light dose (i.e., is favorable in terms of photobleaching and phototoxicity). It is also fast and allows following live or dynamic samples. Its main limitation is that the enhancement of spatial resolution is very limited, typically a factor 2. This is typically the technique of choice for sensitive and dynamic samples and when labeling with special markers is not possible. SIM is discussed in Chapter 4.

Stimulated emission depletion microscopy (STED) is an alternative method that pushes resolution below 100 nm, in three dimensions and multiple colors. STED is based on a confocal illumination where, together with the gaussian illumination beam, a second donut-shaped beam is used to deplete (i.e., turn off) fluorescence in the outer part of the excitation PSF. The resulting PSF is therefore shrunk into a smaller effective profile, improving the lateral resolution. STED offers excellent resolution and can be performed in relatively deep samples in 3D thanks to the confocal illumination. Its main limitation is the very high laser power needed for the depletion donut that results in high photobleaching and phototoxicity. STED is discussed in Chapter 3.

Finally, single molecule localization microscopy (SMLM) is a group of methods all based on the detection of individual molecules and localization of such events. A single molecule obviously emits a symmetric PSF signal, and the high-precision in identifying the position of such a PSF by a fitting procedure allows resolution down to few nanometers in some ideal cases. The wild variety of methods proposed are all based on this principle but vary on how the single molecule's signal is achieved: photo-switching (STORM, GSD), photoactivation (PALM), reversible binding (PAINT, IRIS). These are the methods with the best lateral and axial resolution, but they are typically very slow, limited to TIRF or HiLo illumination, and therefore not suitable for deep samples. SMLM methods are discussed in Chapter 2.

1.4 WHY SUPER-RESOLUTION MICROSCOPY FOR MATERIALS?

The first application of SRM was undoubtedly in the cell biology field. Using the enhanced resolution to solve subcellular structures with nanometer precision led to several breakthroughs in our understanding of cells and sometimes the discovery of completely new cellular structures not even imagined before [33]. However, the potential of SRM to be a powerful tool in fields beyond biology is getting more and more evident, and materials science is a clear case of it. Scientists are synthesizing materials of growing complexity, and there is a strong need for tools able to unveil their structure and function. Often our ability to synthesize molecules and materials is no longer the bottleneck toward materials discovery of an application but rather our ability to characterize properly what is in our vials. Without a proper characterization and understanding of structure-activity relations, it is not possible to rationally design improved materials nor is there a possibility of a proper quality control of the produced formulations. We may argue that the next generation of materials is more and more reminiscent of biological machineries. Therefore, it may need the same analytical tools currently used in structural biology. SRM can fit this picture as well as other techniques. For example, think of the use of cryoEM in structural biology and materials characterization.

Several chapters of this book will be dedicated to highlighting a specific application of nanoscopy in the materials arena, showing unique information that was completely hidden due to the lack of suitable techniques. The great potential of super-resolution microscopy relies on the combination of a few key features: i) the nanometer-scale resolution (obviously!); ii) the specific molecular information coming from labeling; iii) the quantitative nature of many nanoscopy methods; iv) the ability to measure in situ and in operando even soft and delicate materials.

While the enhancement in resolution is an obvious advantage of SRM, it does not justify the use per se of these techniques. Indeed, EM and AFM are still outperforming SRM in terms of absolute obtainable resolution.

However, the combination of resolution with the ability to measure specific molecular entities into a complex mixture (e.g., a multicomponent material) is unique and extremely powerful. Moreover, several nanoscopy techniques allow not only to visualize but quantify features of the structure under examination. This can go as far as literally counting how many molecules are present in a specific area of interest, something hardly accessible with any other technique. If it is considered that this process can often be performed in the real setting of a material (e.g., in liquid, inside a living cell, inside a device), it is clear that SRM has a bright future in the material chemistry field and has the potential to become one of the standard tools in material characterization.

REFERENCES

1. J. R. Lakowicz, ed., *Principles of Fluorescence Spectroscopy* (Springer US, 2006).
2. P. Holzmeister, A. Gietl, and P. Tinnefeld, "Geminate recombination as a photoprotection mechanism for fluorescent dyes," *Angew. Chem. Int. Ed.* **53**, 5685–5688 (2014).
3. A. Baeyer, "Ueber eine neue Klasse von Farbstoffen," *Berichte Dtsch. Chem. Ges.* **4**, 555–558 (1871).
4. J. B. Grimm and L. D. Lavis, "Caveat fluorophore: an insiders' guide to small-molecule fluorescent labels," *Nat. Methods* **19**, 149–158 (2022).
5. M. Cosentino, C. Canale, P. Bianchini, and A. Diaspro, "AFM-STED correlative nanoscopy reveals a dark side in fluorescence microscopy imaging," *Sci. Adv.* **5**, eaav8062 (2019).
6. M. Chalfie, "GFP: lighting up life (Nobel lecture)," *Angew. Chem. Int. Ed.* **48**, 5603–5611 (2009).
7. R. Y. Tsien, "Constructing and exploiting the fluorescent protein paintbox (Nobel lecture)," *Angew. Chem. Int. Ed.* **48**, 5612–5626 (2009).
8. "The Nobel Prize in Chemistry 2008 – Illustrated presentation - NobelPrize.org," https://www.nobelprize.org/prizes/chemistry/2008/illustrated-information/
9. M. Zimmer, "GFP: from jellyfish to the Nobel prize and beyond," *Chem. Soc. Rev.* **38**, 2823 (2009).
10. S. B. Rizvi, S. Ghaderi, M. Keshtgar, and A. M. Seifalian, "Semiconductor quantum dots as fluorescent probes for *in vitro* and *in vivo* bio-molecular and cellular imaging," *Nano Rev.* **1**, 5161 (2010).
11. C. Walkey, E. A. Sykes, and W. C. W. Chan, "Application of semiconductor and metal nanostructures in biology and medicine," *Hematol. Am. Soc. Hematol. Educ. Program* 701–707 (2009).
12. J. Tang, J. Ren, and K. Y. Han, "Fluorescence imaging with tailored light," *Nanophotonics* **8**, 2111–2128 (2019).
13. Y. I. Park, K. T. Lee, Y. D. Suh, and T. Hyeon, "Upconverting nanoparticles: a versatile platform for wide-field two-photon microscopy and multi-modal in vivo imaging," *Chem. Soc. Rev.* **44**, 1302–1317 (2015).

14. E. Abbe, "Beiträge zur Theorie des Mikroskops und der mikroskopischen Wahrnehmung," *Arch. Für Mikrosk. Anat.* **9**, 413–468 (1873).

15. S. Hell and E. H. K. Stelzer, "Properties of a 4Pi confocal fluorescence microscope," *J. Opt. Soc. Am. A* **9**, 2159 (1992).

16. S. W. Hell and J. Wichmann, "Breaking the diffraction resolution limit by stimulated emission: stimulated-emission-depletion fluorescence microscopy," *Opt. Lett.* **19**, 780 (1994).

17. T. A. Klar and S. W. Hell, "Subdiffraction resolution in far-field fluorescence microscopy," *Opt. Lett.* **24**, 954 (1999).

18. M. G. L. Gustafsson, "Surpassing the lateral resolution limit by a factor of two using structured illumination microscopy. SHORT COMMUNICATION," *J. Microsc.* **198**, 82–87 (2000).

19. E. Betzig, G. H. Patterson, R. Sougrat, O. W. Lindwasser, S. Olenych, J. S. Bonifacino, M. W. Davidson, J. Lippincott-Schwartz, and H. F. Hess, "Imaging intracellular fluorescent proteins at nanometer resolution," *Science* **313**, 1642–1645 (2006).

20. S. T. Hess, T. P. K. Girirajan, and M. D. Mason, "Ultra-high resolution imaging by fluorescence photoactivation localization microscopy," *Biophys. J.* **91**, 4258–4272 (2006).

21. M. J. Rust, M. Bates, and X. Zhuang, "Sub-diffraction-limit imaging by stochastic optical reconstruction microscopy (STORM)," *Nat. Methods* **3**, 793–796 (2006).

22. K. Prakash, B. Diederich, R. Heintzmann, and L. Schermelleh, "Super-resolution microscopy: a brief history and new avenues," *Philos. Trans. R. Soc. Math. Phys. Eng. Sci.* **380**, 20210110 (2022).

23. F. Balzarotti, Y. Eilers, K. C. Gwosch, A. H. Gynnå, V. Westphal, F. D. Stefani, J. Elf, and S. W. Hell, "Nanometer resolution imaging and tracking of fluorescent molecules with minimal photon fluxes," *Science* **355**, 606–612 (2017).

24. A. T. Wassie, Y. Zhao, and E. S. Boyden, "Expansion microscopy: principles and uses in biological research," *Nat. Methods* **16**, 33–41 (2019).

25. K. J. A. Martens, S. P. B. Van Beljouw, S. Van Der Els, J. N. A. Vink, S. Baas, G. A. Vogelaar, S. J. J. Brouns, P. Van Baarlen, M. Kleerebezem, and J. Hohlbein, "Visualisation of dCas9 target search in vivo using an open-microscopy framework," *Nat. Commun.* **10**, 3552 (2019).

26. C. Niederauer, M. Seynen, J. Zomerdijk, M. Kamp, and K. A. Ganzinger, "The K2: open-source simultaneous triple-color TIRF microscope for live-cell and single-molecule imaging," *HardwareX* **13**, e00404 (2023).

27. A. Auer, T. Schlichthaerle, J. B. Woehrstein, F. Schueder, M. T. Strauss, H. Grabmayr, and R. Jungmann, "Nanometer-scale multiplexed super-resolution imaging with an economic 3D-DNA-PAINT microscope," *ChemPhysChem* **19**, 3024–3034 (2018).

28. J. S. H. Danial, J. Y. L. Lam, Y. Wu, M. Woolley, M. R. Cheetham, D. Emin, and D. Klenerman, "Constructing a cost-efficient, high-throughput and high-quality single molecule localization microscope for super resolution imaging," 1–50 (n.d.).

29. B. Diederich, P. Then, A. Jügler, R. Förster, and R. Heintzmann, "cellSTORM—Cost-effective super-resolution on a cellphone using dSTORM," *PLOS ONE* **14**, e0209827 (2019).

30. U. Boehm, G. Nelson, C. M. Brown, S. Bagley, P. Bajcsy, J. Bischof, A. Dauphin, I. M. Dobbie, J. E. Eriksson, O. Faklaris, J. Fernandez-Rodriguez, A. Ferrand, L. Gelman, A. Gheisari, H. Hartmann, C. Kukat, A. Laude, M. Mitkovski, S. Munck, A. J. North, T. M. Rasse, U. Resch-Genger, L. C. Schuetz, A. Seitz, C. Strambio-De-Castillia, J. R. Swedlow, and R. Nitschke, "QUAREP-LiMi: a community endeavor to advance quality assessment and reproducibility in light microscopy," *Nat. Methods* **18**, 1423–1426 (2021).

31. S. Pujals, N. Feiner-Gracia, P. Delcanale, I. Voets, and L. Albertazzi, "Super-resolution microscopy as a powerful tool to study complex synthetic materials," *Nat. Rev. Chem.* **3**, 68–84 (2019).

32. L. Schermelleh, A. Ferrand, T. Huser, C. Eggeling, M. Sauer, O. Biehlmaier, and G. P. C. Drummen, "Super-resolution microscopy demystified," *Nat. Cell Biol.* **21**, 72–84 (2019).

33. S. J. Sahl, S. W. Hell, and S. Jakobs, "Fluorescence nanoscopy in cell biology," *Nat. Rev. Mol. Cell Biol.* **18**, 685–701 (2017).

CHAPTER 2

Localization Microscopy

Lorenzo Albertazzi and Peter Zijlstra

Eindhoven University of Technology, Eindhoven, The Netherlands

2.1 INTRODUCTION

The resolution of far-field optical microscopies has traditionally been limited to several hundred nanometers due to the diffraction limit of light. Nevertheless, optical microscopy has remained immensely popular under biologists, chemists, physicists, and materials scientists alike because it provides the distinct advantages of (1) being relatively non-invasive and (2) chemically specific. The first advantage provides the opportunity to observe fragile samples such as live cells and soft materials in their natural environment (typically an aqueous environment). The latter advantage is achieved by chemically specific labeling of structures using the methods described in Chapter 1. This enables the observation of specific structures in an otherwise complex sample, while multi-color labeling approaches enable the study of multiple structures simultaneously.

Recently several approaches have been introduced that circumvent the diffraction limit of light (note that "beating" the diffraction limit is not possible, it is a fundamental limit that follows from Maxwell's equations). These approaches are collectively called super-resolution fluorescence microscopy. One class of methods circumvents the diffraction limit by using engineered illumination patterns in a microscope. Examples of these methods include microscopy based on stimulated emission depletion (STED, Chapter 3) and structured illumination microscopy (SIM, Chapter 4). A second class of methods employs the fact that individual emitters can be switched on and off repeatedly, while their camera image can be used to determine the location of the emitter with a precision that is better than the microscope's resolution.

DOI: 10.1201/9781003220688-2

In this chapter we focus on the latter method. We first describe principles of localization microscopy, where we will outline the underlying concepts and the workflow in a typical experiment. We then highlight approaches to switch individual dyes on and off, which is arguably the most important aspect of this type of super-resolution microscopy. We then highlight methods to go from 2D imaging to 3D imaging, and from single-color to multi-color microscopy. We finish the chapter with application examples in the area of materials science.

2.2 PRINCIPLES OF 2D LOCALIZATION MICROSCOPY

Single-molecule localization microscopy (SMLM) is performed in a standard fluorescence microscope (e.g., the TIRF microscope described in Chapter 1). This technique has become an attractive method because it employs standard setups and such hardware is already present in a large number of laboratories and companies world-wide. To turn a standard TIRF fluorescence experiment into a super-resolution localization experiment requires three modifications of the workflow:

- The first modification is to employ samples in which only a small part of the fluorescent dyes is in an emitting state, whereas the majority of emitters are in a dark state. This generates a camera image that contains several spots that represent the location of the emitting fluorescent dyes, while the remaining dyes are invisible because they remain dark. Several approaches have been developed that induce stochastic switching of the dyes between the bright and the dark state. This then allows for the localization of a different subset of emitters in a series of camera frames. Such methods to switch the fluorescence state of a dye are introduced in Section 2.4. This implies that the TIRF setup needs to be equipped with powerful lasers and a sensitive camera to enable the detection of a single molecule.

- The second modification is to introduce post-imaging algorithms that fit the camera image of a single emitter, typically with a Gaussian profile, and thereby determines the most likely location of the underlying molecule. The precision of this localization depends on the shape of the point spread function of the microscopy, the signal-to-noise ratio of the images, and fitting algorithms used. These aspects are discussed below in Section 2.3.

- Finally, accumulating all localizations across a number of frames yields the location of a large number of fluorescent dyes, as shown in Figure 2.1. If these dyes are chemically attached to a specific structure (e.g., a polymer or a particle) the geometry of the underlying structure can be reconstructed. Note that an SMLM image contains information on both the overall structure of the sample and the number and position of the dye molecules on the structure. This enables quantitative analysis on the molecular level using the molecular counting discussed in Chapter 8.

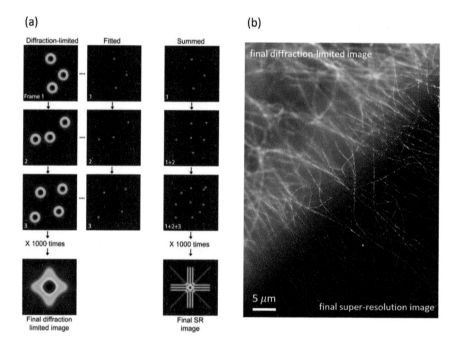

FIGURE 2.1   (a) Illustration of the workflow in a typical super-resolution localization experiment. Camera frames (left column) of a fluorescently labeled sample are recorded, and the location of single dye molecules is determined by fitting the point spread function. The fluorescent labels are chosen in such a way that they stochastically switch between bright and dark states, enabling the localization of a different subset of emitters in subsequent frames. Accumulating all localizations results in a super-resolution reconstruction of the underlying structure (bottom right) that reveals substantially more detail than the diffraction limited image (bottom left). (b) Experimental result showing the final diffraction limited image of supra-molecular polymers, compared to the final super-resolution image obtained with the workflow in (a). ([a] Adapted with permission from [1]. Copyright 2014 American Chemical Society.)

2.3 LOCALIZING THE EMITTER BY FITTING THE POINT SPREAD FUNCTION (PSF)

2.3.1 Shape of the PSF

A crucial step of SMLM is the fitting of the PSF of single fluorescent molecules. The image that a single emitter (e.g., a dye molecule) creates on a camera depends on both the properties of the emitter and the properties of the microscope. We will assume that the fluorescent dye is attached to the structure via a flexible linker, and thus rotates freely. For small molecules the rotational Brownian motion occurs on timescales of nanoseconds (depending on the size of the fluorescent dye and the stiffness of the linker). This implies that the dye will explore all orientations within a single camera frame because these are typically captured with a millisecond integration time. If we further assume that the refractive index in the environment of the dye is homogeneous and isotropic, the fluorescent dye can be approximated as an isotropic emitter that emits a spherical wave.

This spherical wave then travels away from the fluorescent dye, part of it being captured by the objective lens (see Figure 2.2). We consider a very simple microscope, consisting of only one lens with focal length f, playing the role of both objective and tube lenses focusing the light from the dye on the camera. The collection lens has a finite diameter (D) and will collect only part of the emitted light, while the largest part of the emission will be lost. This accurate simplification results in a situation where the image of the emitter equals the focusing of light that has diffracted from a circular aperture. In a microscope the lens acts as both the focusing element and

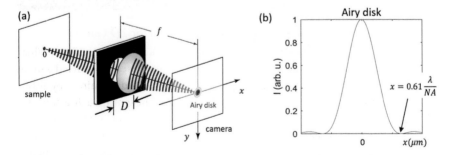

FIGURE 2.2   In an optical microscope (here represented with a single lens) spherical waves emitted from the dye are diffracted by the lens aperture and focused onto the camera. Interference of the emitted waves in the camera plane generates an Airy disk. (Adapted with permission from [2]. Copyright 2019 John Wiley & Sons. (b) Cross section of the Airy disk with the first minimum indicated.)

the aperture because only the rays reaching the lens will be transmitted to the camera plane.

The emission of the dye molecule will thus diffract off the aperture before being focused onto the camera. As a result, the point source appears on the screen not as a point, but as an Airy disk with a finite size [2]. The size of the Airy disk, and thus the resolution of an optical microscope, can be quantified by the disk's first minimum that is given by $x = 0.61\ \lambda/\text{NA}$, which is the Rayleigh diffraction limit. This indicates that the resolution of an optical microscope depends on the wavelength of the emitted light λ and the numerical aperture of the objective lens ($\text{NA} = n \sin \theta = nD/2f$, where n is the refractive index of the medium between lens and sample). Shorter wavelengths and larger apertures (that capture more of the emitted light) improve the resolution. Nevertheless, the resolution of an optical microscope is most often quantified via the Abbe diffraction limit, which is a little more optimistic ($x = \lambda/2\text{NA}$).

The simplified analysis of the PSF above is valid in the limit of low numerical aperture, but in reality, the numerical aperture of a good oil immersion lens is 1.4–1.5. The result is that the PSF effectively approaches a more Gaussian shape, with smaller shoulders at the location of the first Airy-rings [3]. These small shoulders are most often not clearly resolved in a real experiment because of noise sources, most notably camera dark noise and shot noise. This is illustrated in Figure 2.3 that shows the camera image of a freely rotating dye molecule (ATTO655 in this case) that is imaged on a camera using an objective lens with a numerical aperture of 1.49. The 1D cross section shows a near-Gaussian distribution of photons where the Airy rings are less visible (Figure 2.3, right).

For this reason, localization microscopy mostly assumes a 2D Gaussian distribution of photons on the camera. This allows for 2D Gaussian of the PSFs that is conceptually simple and computationally efficient, despite the fact that it lacks a physical basis. Note that experiments where the fluorophore is not free to rotate require a very different approach, as was outlined recently by Stallinga and Rieger [3] and is further discussed in Chapter 6.

2.3.2 Fitting the PSF

The concept of localization microscopy is based on the ability to fit the PSF of a single molecule and pinpoint its location with a precision that is higher than the resolution of the microscope. As described above, most often the PSF is approximated by a 2D Gaussian distribution of photons

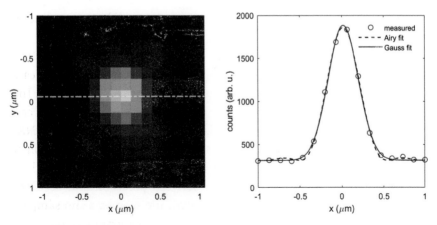

FIGURE 2.3   Measured microscopy image of a single emitter (ATTO655) imaged on a camera with an objective lens with NA = 1.49. The right graph compares the measured cross section to the shape of the Airy disk (the expected PSF in the limit of low NA) and a Gaussian distribution.

whose center is determined using software. The algorithms used in the software vary but are mostly based on statistics: one writes down a model for how the intensity $I(x, y)$ on a pixel at (x, y) depends on the coordinates of the fluorophore, its photon emission rate, the background in the experiment and, in most advanced analysis, other parameters such as aberrations or the amount of defocus. The parameters are varied to find the values that give the best 'fit' to the data, where the goodness of the fit is typically measured by either the maximum-likelihood (ML) criterion or the least-squares criterion (LS). Once the best fit is found, coordinates are inferred from the fit parameters [4, 5].

A localization algorithm without bias has a precision that is limited by the Cramér-Rao lower bound [6–9]. ML estimation requires a model for the PSF as well as a model for the noise distribution for the number of photons in a pixel. Using these two models, a likelihood function is constructed that measures the mismatch between the spot and the PSF model, accounting for the variance of the number of photons. A key advantage of MLE is that it is theoretically proven to be the optimal fitting procedure. This means that it achieves the best-possible precision, given by the Cramér-Rao lower bound, when the estimator is unbiased and the image is formed by a large total number of photons. For high photon counts, LS and ML estimators perform equally well, but at lower photon counts ML estimators will significantly outperform LS fitting [4, 10, 11]. Note that all the above-mentioned considerations assume the PSF model is correct;

a wrong PSF model will lead to a bias or a lower precision. A more detailed description of fitting algorithms can be found in Chapter 6.

2.3.3 Localization Precision and Accuracy

Owing to noise, localization algorithms always make errors. These errors have random and systematic parts as measured by the variance and the bias. As explained by Braeckmans et al. [12], these errors can be quantified using the notions of localization precision (quantifying the variance) and localization accuracy (quantifying the bias).

Figure 2.4 illustrates the difference between precision and accuracy, where precision indicates the spread in localizations when a single fluorophore is localized multiple times in an experiment. Due to noise, each localization will be at a slightly different location, even though the fluorophore has not moved. Localization accuracy on the other hand is related to the bias of the cloud of localizations, in other words it quantifies the difference between the true location of the molecule and the location that is found in an experiment (i.e., the center of the cloud of localizations).

Starting with the localization precision, its ultimate value is limited by the Cramér-Rao bound. The CRB expresses an upper bound on the precision of unbiased estimators. This limit mainly depends on the signal-to-noise ratio of the camera images, where a higher signal and lower noise yield more precise localizations. In general, the localization precision along one of the camera axes is given by

$$\sigma_x \geq \sigma_0 / \sqrt{N}, \tag{2.1}$$

where σ_0 is the standard deviation of the PSF that is related to its full width at half maximum by $\text{FWHM} = 2\sqrt{2\ln 2}\sigma_0$, and N is the integrated number of photons in the PSF. Typical values are $\sigma_0 = 120$ nm and $N = 10^4$ photons,

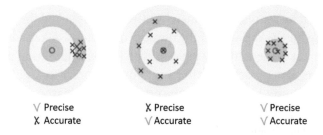

√ Precise   X Precise   √ Precise
X Accurate   √ Accurate   √ Accurate

FIGURE 2.4   Quantification of localization microscopy data using localization precision and accuracy. (Adapted with permission from [13]. Copyright 2021 Springer Nature.)

which predicts a precision of $\sigma_x = 1.2$ nm. However, this equation neglects any other noise sources apart from shot noise. Other sources of noise typically present are the camera's read and dark noise, background signals, and finite camera pixel-size. These noise sources have been accounted for in more complex models, for example [4]:

$$\sigma_x \geq \frac{\sigma_a}{\sqrt{N}} \sqrt{\left(\frac{16}{9} + \frac{8\pi\sigma_a^2 b^2}{Na^2} \right)}. \tag{2.2}$$

Herein $\sigma_a^2 = \sigma_0^2 + a^2/12$ with a the pixel size, whereas b^2 is the expected number of background photons per pixel. Equation (2.2) provides a theoretical estimate of the localization precision when using ML estimators for the PSF fitting and can be used if typical imaging parameters are known. Note that, in the limit of infinitely small pixel size ($a \to 0$) and a noise-free background ($b \to 0$) we recover Eq. (2.1) with an additional factor of $\sqrt{\frac{16}{9}}$ that comes from the statistics of the noise in the PSF and distinguishes it from the original equation by Thomson that systematically overestimates the localization precision [14].

Figure 2.5 shows the estimated localization precision, where the simplified shot noise limit (Eq. (2.2.1)) is plotted as the dashed line for reference.

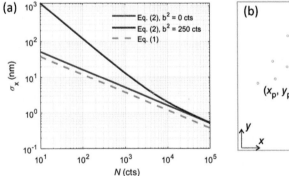
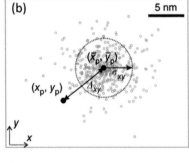

FIGURE 2.5   (a) Calculated localization precision. The yellow dashed line assumes that only shot noise affects localization (Eq. (2.2.1)). The blue and red lines are plots of Eq. (2.2) for a PSF with size $\sigma_0 = 120$ nm and a pixel size of $a = 100$ nm. (b) measured localization precision showing a large number of localizations of the same emitter. The localization precision is indicated by σ_{xy}, whereas the localization accuracy is indicated by Δ_{xy}. (Adapted with permission from [12]. Copyright 2014 Springer Nature.)

The effect of the finite pixel size being rather modest, the effect of background is substantially larger, particularly for dim fluorophores. This means that background is a key parameter in these experiments and should be minimized at all costs. Having said that, the background noise can never be absolutely zero because of noise sources in the camera.

In contrast to the above theoretical estimate of the localization precision, it can also be measured directly. This is typically done by repeatedly localizing a small emitter such as a single dye molecule or a nanometer-sized fluorescent bead [12]. By repeatedly localizing the emitter and accumulating the localizations in a scatter plot such as in Figure 2.5b, the standard deviation of the cloud of localizations provides the measured localization precision. Because not all aberrations can be included in the theoretical estimate of the localization precision, such measurement is a more robust approach.

Apart from the variance introduced by the localization algorithms they may also induce a bias. This results in an inaccurate localization, i.e., the cloud of localizations is shifted away from the true position on the molecule. Because the true position of the molecule is never known (it requires a second independent method of localization) the accuracy is most often unknown.

In addition to inaccuracies induced by the optics or the localization algorithm, also the size of the fluorescent label can introduce inaccuracies: although organic dyes are approximately 1 nm in size, in biological experiments they are often conjugated to the underlying structure using e.g., antibodies that are ~10 nm in size. The size of the label then becomes comparable to the localization precision, so instead of reconstructing the underlying structure one may actually reconstruct the position of the label, also referred as linkage error [15]. Another source of inaccuracy may be the movement of the sample [16] or of the molecule itself [17] during image acquisition. Note that the acquisition of SMLM images may take minutes or even hours, so sample drift plays an important role and has to be minimized and corrected to achieve an optimum resolution.

2.3.4 Image Reconstruction

The underlying structure of the sample is recovered after localizing all fluorophores in the sequence of camera images and accumulating their location in a reconstruction (see Figure 2.1). The "resolution" of the final image is then rather difficult to quantify because it depends on both the localization precision of a single fluorophore, as well as the density of labels in the sample. The most faithful reconstruction will require (nearly) all dyes to be switched on and off at least once. As further described in Section 2.4, the switching is often a stochastic process that requires a large number of frames (sometimes >10^4 frames) to be captured before the underlying structure can be faithfully reconstructed.

For continuous structures, the effect of the total number of localized fluorophores is illustrated in Figure 2.6 where simulations are used to resolve circles of different sizes with a certain number of total localizations. It can clearly be seen that small circles (i.e., high spatial frequencies) are only resolved when the number of localizations is higher than some critical value. This critical value can be quantified using the Nyquist-Shannon sampling criterion, which states that structural details smaller than twice that of the average label-to-label distance cannot be reliably resolved. Thus, for a two-dimensional image having spatial features of size ρ, the minimum required molecular density of localized fluorescent probes necessary to meet the Nyquist criterion is $(2/\rho)^2$, or more generally stated $(2/\rho)^D$ where D is the dimensionality of the image.

Therefore, for example, to achieve 20 nanometer resolution in two dimensions, one fluorophore has to be positioned at least every 10 nanometers, and an extremely high molecular density of around 10,000 molecules per μm^2 is required. For super-resolution in three dimensions, 20 nanometer resolution requires about 10^6 fluorophores per μm^3. In general, a sufficient density of fluorescent probes must be present in order to fully map the fine details of a labeled structure. By the same criterion, a sufficient fraction of these probes must be localized during the imaging process.

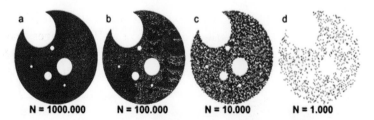

FIGURE 2.6 Effect of labeling density on image reconstruction. This simulation shows a sample structure with different length scales (i.e., circles with different sizes). The small circles are only resolved if the number of localizations exceeds the Nyquist-Shannon sampling criterion. (Adapted with permission from [18]. Copyright 2015 Springer Nature.)

2.4 METHODS FOR SWITCHING DYES

As clearly emerges from the previous sections, in order to achieve a super-resolved image, it is necessary to have a high density of emitters but at the same time be able to localize individual ones, i.e., to not have overlapping emitters. As a consequence, only an extremely low fraction of the emitters should be bright at a given time. Therefore, controlling the fluorescent on- and off-states of the fluorophores is critical to a successful experiment. Since the 1990s, steady developments in the areas of protein-based and synthetic fluorophores have enabled temporal control of the fluorescence with a range of approaches that are all viable and can, e.g., be selected based on the goal of the experiment and the type of sample. The methods were proposed in the 2000s in the wake of a general emergence of optical super-resolution microscopy methods; the most used approaches are displayed in Figure 2.7.

2.4.1 Photoactivation (PALM)

One common way to control the on-off behavior of dyes is using light. Specific wavelengths may trigger photophysical and photochemical processes that turn on or turn off emitters, allowing photocontrol. Photoactivation (switching from an off-state to an on-state upon illumination at a specific wavelength), photoswitching (i.e., reversible switching between an on-state and an off-state upon illumination at two different wavelengths), and photoconversion (i.e., conversion of fluorescent state from one color to another upon illumination at a specific wavelength) have been the most widely used strategies.

Fluorescence photoactivated localization microscopy (PALM, Figure 2.7a) was initially demonstrated using fluorescent proteins that can be activated

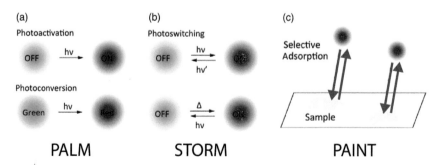

FIGURE 2.7 Methods to switch the state of a fluorescent dye for super-resolution localization microscopy. (a) photoactivation or conversion, (b) photoswitching, (c) switching induced by reversible and selective interactions.

by UV illumination [19]. The development of PALM as an imaging method was therefore largely prompted by the discovery of new species and the engineering of mutants of fluorescent proteins displaying a controllable photochromism, such as photo-activatable green-fluorescent protein (GFP). In addition to the use of fluorescent proteins, synthetic dyes have also been developed for PALM microscopy, such as caged fluoresceins, rhodamines, or cyanines. The advantages of using synthetic dyes are their (generally) higher brightness and smaller size, which provide higher localization precision and the ability for denser labeling. In a standard PALM experiment dyes are photoactivated sparsely, so that only a small subset of molecules is activated at the same time. These molecules are localized and then irreversibly bleached. The last bleaching step is necessary to turn off the dyes after localization to leave "space" for a new dye to activate. This represents an advantage and at the same time a disadvantage of PALM. On one side the fluorophores are activated only once, allowing for precise molecular counting, on the other this limits the sampling of the emitters, reducing reconstruction and precision. PALM is generally a versatile method that does not require specific buffers or conditions and is compatible with fluorescent proteins and live cell imaging, where it is often combined with single-particle tracking following activation (SPT-PALM). Compared to other SMLM methods, PALM often suffers from lower resolution as the probes used are less bright, and multicolor workflows are very complex due to the difficulty of combining multiple probes with separate emission and activation wavelengths.

2.4.2 Photoswitching (STORM)

While PALM uses irreversible photoactivation, reversible photoswitching can also be used. In this case, molecules able to switch between a bright "on" state and a dark "off" state are necessary. The on-off transitions can be catalyzed by a specific laser wavelength – one color turns on, another color turns off – or a combination of light-induced and spontaneous thermal recovery (see Figure 2.7b). This can be achieved through many processes but the most common is the rapid photoreduction of fluorophores. Upon illumination with an excitation beam in the presence of appropriate reducing reagents, some organic fluorophores undergo a light-triggered reduction, trapping the fluorophore in a metastable dark state which can act as the off-state. Based on this mechanism, the fluorescent state of cyanine dyes, Alexa dyes, and Atto dyes can be switched reversibly between on- and off-states.

This approach is called Stochastic Optical Reconstruction Microscopy (STORM). The original implementation of STORM [20] again made use of a pair of cyanine dyes: a photo-switchable "reporter" fluorophore was cycled between fluorescent and dark states, which was facilitated by an "activator" dye that was connected via flexible linker [21]. For the Cy3-Cy5 pair for example, upon illumination with a red laser, the Cy5 was initially fluorescent and then quickly switched into a non-fluorescent, dark state. A brief exposure to a green laser pulse led to stochastic reactivation of a subset of the dyes back to the fluorescent state. These reporters typically could be switched on and off for hundreds of cycles before permanently photobleaching, increasing the reconstruction and precision.

A different implementation of STORM only requires the use of a single type of fluorophore and is therefore dubbed "direct" STORM. Early work [22, 23] showed that commercially available unmodified carbocyanine dyes such as Cy5 (usually excited at 633 nm) can be used as efficient reversible single-molecule optical switch, whose fluorescent state after apparent photobleaching can be restored. The rate for the transition from the on-state to the off-state is controlled by the concentration of a reducing thiol reagent in the buffer, and the irradiation intensity. Later studies suggested that the photoconversion product is a thiol–cyanine adduct in which covalent attachment of the thiol to the cyanine disrupts the original conjugated π-electron system of the dye and renders it dark [24]. The rate of the transition back into the on-state depends on the thermal stability of the adduct and can be controlled by the concentration of molecular oxygen or irradiation with low-intensity UV light. The result is that the molecule returns to the ground state, and fluorescence is recovered.

STORM is heavily used for imaging biological and synthetic structures thanks to its good resolution, ease of implementation, and multicolor ability. A key aspect to take into account is that the performance is strongly dependent on the buffer conditions and buffer optimization is a crucial step. Standard STORM buffers contain a reducing agent (typically a thiol such as MEA) and an oxygen scavenging system to prevent excessive photobleaching (typically a GLOX/Catalase enzymatic system). Each dye prefers slightly different conditions, making sample preparation more challenging, especially for multicolor imaging.

An important experimental parameter for high-quality dSTORM super-resolution imaging is an appropriate ratio of the on- and off-switching rates. The higher the fluorophore density, the higher must be the fraction of fluorophores in the off-state to ensure that each fluorophore is recorded as a single molecule at a given time. This implies that the off-switching rate should be much larger than the on-switching rate to guarantee sufficiently low spot densities per imaging frame. These switching rates are not only controlled by the laser power densities, but also by the buffer composition and temperature of the sample. This makes it challenging to control the rates in a broad range of conditions, which has sparked the development of switching-mechanisms that do not rely on chemical reactions but rather on reversible adsorption and desorption of the fluorophore, see Figure 2.7c.

2.4.3 Switching by Reversible Interactions

While PALM and STORM are based on the control of the emitting state of a fluorophore, a switching mechanism can be envisioned based on the reversible adsorption and desorption of fluorescent probes that are always emitting. The principle of this approach is based on the notion that an unbound fluorophore in solution diffuses so fast that it only provides a (low) number of diffuse background photons, spread across the whole image. A fluorophore that is shortly bound to the sample, however, is immobilized and gives rise to a PSF that can be localized similarly to PLAM and STORM. The bound state of the fluorophore is therefore representative of the on-state, whereas the unbound state represents the "off-state." In contrast to PALM and STORM, note that the fluorophores here still emit in the apparent "off-state," but do not generate a PSF because they diffuse too fast. This family of methods is generally referred to as PAINT, point-accumulation for imaging nanoscale topography.

Therefore, the key of these methods is the mechanism of reversible interactions with the sample that provides a straightforward means to control the binding rate. Ideally, a probe has a high tendency to be in the bound state (i.e., high k_{on}) and a tuned lifetime of the bound state (tuned k_{off}). The bound lifetime should be long enough to accumulate enough photons for a precise localization but short enough to "leave space" for the binding and detection of adjacent sites. The binding rate can be controlled by the intrinsic affinity of the probe for the sample, the concentration of the probe in solution, and the buffer conditions as they may influence the k_{on} and k_{off} of the probe. Typical conditions employ probes with a K_D ($K_D = k_{off}/k_{on}$) around 100 nM and pM-low nM concentration of probes in solution.

The most straightforward way to control both rates is by using DNA as a mediator, see Figure 2.8a. This method is called DNA-mediated point-accumulation for imaging nanoscale topography, or DNA-PAINT.

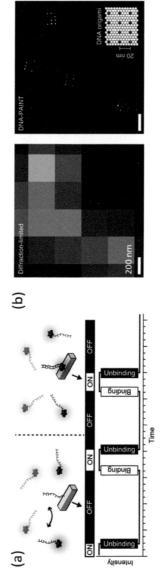

FIGURE 2.8 (a) Illustration of the concept of localization microscopy based on transient molecular interactions mediated by DNA. The short binding events are detected in the microscope as single-molecule "blinks," as illustrated in the timetrace below. These "blinks" can be fitted with a PSF model to yield a super-resolution localization of the site at which the interaction took place. (b) Example of an experiment where the final diffraction-limited image for a DNA-origami structure with 12 docking strands is compared to final super-resolution image. (Adapted with permission from [25]. Copyright 2017 Springer Nature.)

The fluorophore is conjugated to a short DNA strand, often referred to as the imager strand. The underlying structure to be resolved is then labeled with complementary strands (docking strands) that specifically bind the imager. The complementary sequence can be synthetically controlled with single-nucleotide resolution.

Typically, association rates for DNA hybridization are 10^6 $M^{-1}s^{-1}$ [25] that depend on the buffer composition and DNA-sequence [26]. This provides a robust mechanism to control the binding rate where typical imager concentrations of pico- to nanomolar are used to control the number of bound imagers in the field-of-view of the microscope. An additional advantage comes from the fact that this approach decouples the "blinking" behavior (now controlled by DNA) from the intrinsic properties of the fluorophores. This means that essentially all fluorophores can be used in DNA-PAINT, rather than a small subset that exhibits a special photophysics. One of the main advantages of DNA-PAINT is that the blinking kinetics is now decoupled from the fluorophore itself, enabling the use of the best and brightest fluorophores available as long as they can be conjugated to the imager DNA-strand. This results in localization precisions of a few nanometers (see Figure 2.8b), where the resolution of the reconstruction is, in practice, limited by drift of the microscope [27].

While DNA-PAINT is still the most used PAINT method, several other reversible interactions can be used. Recently, many probes using hydrophobic dyes [28], peptide- and protein-protein interactions [29], and carbohydrates [30] have been employed for PAINT microscopy.

2.5 QUANTIFICATION APPROACHES

Once localizations have been collected in a measurement, the resulting maps can be quantified. Note that the raw data produced by SMLM is not a classical pixel-based image but rather a list of molecular events, with their x, y, z, t coordinates and their fitting parameters. These data sets are very rich in information on the molecular level, but extracting meaningful properties is far from trivial. One challenge is that one fluorophore can give multiple localizations; therefore, the translation of localizations (i.e., frames in which a fluorophore was in the bright state) to actual molecular sites (i.e., the number and localization of biomolecules or active sites on a sample) is not a trivial task. The blinking kinetics of fluorophores originate from a random process, so the number of blinks is not straightforwardly converted to the number of underlying molecules. Approaches have been developed to perform this conversion to quantitatively analyze localization maps in terms of the number of active sites [31, 32] and their distribution (e.g., randomly distributed, clustered, or otherwise organized) [33–35].

When active sites are sparse and spaced by more than twice the localization precision of the measurement (as in Figure 2.9a) the active sites are clearly resolved as groups of localizations. Each group of localizations can then be assigned to an active site and visualized using, e.g., Voronoi tessellation (see Figure 2.9b). This allows for the direct quantification of the number and distribution of the underlying sites. The degree of randomness in the distribution of sites can then be extracted, which provides important information on, e.g., the distribution of receptors on the cell surface [36] or the distribution of active sites on a catalyst structure [37]. Quantification of the degree of randomness can be achieved using, e.g., nearest neighbor analysis, where pairwise distances between nearest neighbors are compared to the expected distribution from the random 2D Poisson distribution [35, 38]. Alternatively, pair correlation functions or Ripley's K-function can be employed to quantify the degree of randomness [34, 35].

When active sites are spaced by less than the localization precision, a different approach to quantification has to be taken. This regime corresponds to the one illustrated in Figure 2.9c, where direct counting and localization of each active site is impossible. Approaches have been developed to extract

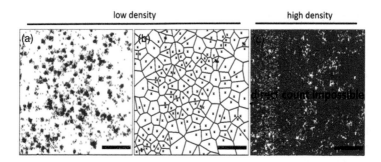

FIGURE 2.9 Illustration of two extreme regimes in which localization microscopy can be performed: at densities where the target molecules are separated by more than the localization precision the underlying molecules can be distinguished by eye from their cloud of localizations. This allows for direct quantification of the number and spatial distribution of the underlying sites as shown in (b). For high densities of target sites other approaches based on statistics are needed. (Adapted with permission under a Creative Commons CC-BY license from [36]. Copyright 2023 The Authors.)

the local density of active sites, where the number of sites can be retrieved but not their exact location because localization clouds overlap. This requires precise knowledge of the blinking behavior of a fluorophore, in other words, knowledge of the average number of bright states per fluorophore.

The first step in such quantification is to group neighboring frames in which the fluorophore was bright, thereby converting localizations to blinking events (one event can result in multiple localizations if the blink lasts longer than one frame). The next step is to relate the number of blinks to the number of underlying sites. For PALM and STORM the photophysics of a fluorophore is often quantified by assuming a four-state model: starting from a dark state, the fluorophore is activated (in PALM by a second laser wavelength, in STORM by a chemical conversion) and becomes bright. From the bright state, the fluorophore has two options: transition to a temporary dark state that returns back to the bright state (blinking) and irreversible photobleaching and permanent loss of emission. Mathematical models [39, 40] then relate the number of measured blinks in a certain area to the number of underlying fluorophores. However, it is important to keep in mind that the photophysical transitions of fluorophores are influenced by their nano-environment, necessitating the calibration of the blinking behavior in a controlled experiment on isolated fluorophores.

Above we have described quantification approaches for very sparse and very dense samples. Intermediate densities often require a hybrid approach where nearest neighbor analysis or pair correlation analysis is combined with statistical approaches to correct for the multiple blinking of fluorophores [41]. A more detailed discussion on quantification of PALM and STORM data can be found in Chapter 8.

A drawback of the above-discussed approaches is the difficulty in controlling the blinking behavior of fluorophores that depends on the fluorophore itself, the linkers used for conjugation, and the local environment. An approach that overcomes this is DNA-PAINT, where the interaction kinetics are dictated by the DNA transient interactions. Rather than relying on the four-state model described above, DNA-PAINT can be described by a two-state model involving the bound- and unbound states. Quantification by DNA-PAINT is often dubbed quantitative PAINT (qPAINT) and relies on precise knowledge of the association rate k_{on} (units $M^{-1}s^{-1}$), see Figure 2.10a. The association rate depends on the sequence and local environment of the docking strand, and is often calibrated in a control experiment on, e.g., DNA-origami structures with a single binding site [26]. Once k_{on} is known, the number of underlying sites can be extracted

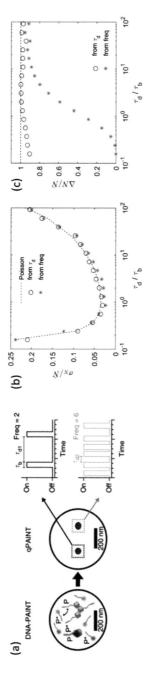

FIGURE 2.10 (a) Illustration of molecular counting using DNA-PAINT. The reproducible single-molecule binding kinetics of DNA-hybridization results in a timetrace where the number of events depends on the number of underlying binding sites. (b) Simulated counting precision σ_N (normalized to the number of binding sites $\bar{N}$). The Poisson limited precision is compared to counting approaches based on the statistics of the dark times (red open circles) and compared to a naïve analysis of the frequency of events, where $f = Nk_{on}c_{img}$. (c) Simulated counting accuracy ΔN (normalized to the number of binding sites $\bar{N}$). The known number of binding sites is compared to the counted number of sites based on the dark times and the frequency of events. ([a] Adapted with permission from [32]. Copyright 2016 Springer Nature; [b] and [c] Reproduced with permission under a Creative Commons CC BY-NC 3.0 License from [42]. Copyright 2020 The Authors.)

directly using $\tau_d = (Nk_{on}c_{img})^{-1}$, where the dark-time τ_d (units seconds) is the average time between two consecutive binding events, N is the number of underlying sites, and c_{img} is the imager concentration.

The average dark-time is often determined by measuring a large number of events and fitting the distribution of dark-times with a single-exponential function $f(t) = (\exp - t/\tau_d)$ to extract the mean τ_d. The degree of quantification obtainable can again be divided into counting precision and accuracy, as we did for the localization precision and accuracy in Section 2.3.3. If the number of sites in a region estimated n times by independent experiments, the counting precision describes the spread of these estimates around its mean value $\bar{N}$, commonly expressed in terms of a standard deviation σ_N. The counting precision is then essentially determined by the number of detected dark times. The counting accuracy on the other hand describes to what degree the mean estimated number of sites $\bar{N}$ deviates from the true number of sites N. These deviations mainly arise when binding events overlap in time, resulting in undercounting. These counting inaccuracies are given by $\Delta N = \bar{N} - N$.

The counting precision and accuracy can be analyzed as a function of the ratio between mean dark- and bright times τ_d/τ_b [42], which can be expressed in the experimental parameters as

$$\frac{\tau_d}{\tau_b} = \frac{k_{off}}{Nc_{img}k_{on}} \qquad (2.3)$$

This ratio between dark- and bright times can be experimentally tuned by simply changing the imager strand concentration. Figure 2.10b shows the normalized counting precision (coefficient of variation), where a low precision is observed for low and high τ_d/τ_b ratios with an optimum at $\tau_d/\tau_b \sim 1$. For large ratios the number of events per timetrace is limited (e.g., due to a low c_{img}), resulting in imprecise counting. For small ratios on the other hand, the large binding frequency results in a fraction of events overlapping in time. Such double events cannot be identified reliably in an experiment, resulting in missed events and thus a reduced total number of events per time trace for decreasing ratios.

The counting accuracy is shown in Figure 2.10c, which shows that an analysis based on the dark times provides a counting accuracy of >90% across a large range of τ_d/τ_b. It is interesting to compare the dark-time analysis to a more naïve approach based on the frequency of events: the latter produces consistent undercounting because each double event affects the measured frequency directly. The accuracy therefore substantially decreases for $\tau_d/\tau_b < 10$ due to the increasing probability of double events. When the mean dark time is considered, the counting accuracy is substantially higher over a larger range of τ_d/τ_b because double events reduce the number of detected dark times (as captured by σ_N) but do not affect their mean. So although the counting precision reduces when a substantial number of events overlap, the counting accuracy is robust against double events.

2.6 APPLICATION TO MATERIALS

Until the 2010s the use of super-resolution microscopy was largely restricted to the biological sciences due to its ability to resolve sub-cellular structures with unprecedented resolution. In eukaryotic cells, many well-characterized structures such as the cytoskeleton and other protein scaffolds have been reexamined with super-resolution fluorescence microscopy. These proof-of-principle systems not only demonstrated the resolving power of the technique but also gave a glimpse of its potential for novel discoveries. The sub-diffraction-limit imaging of the axon cytoskeleton is a striking example [43]. Also, the distribution of membrane receptors in eukaryotic and bacterial cells have been widely investigated. Since the 2010s localization microscopy is increasingly applied in the physical and materials sciences, which we highlight in this section by zooming in on applications in polymer science, nanomedicine, biosensing, and catalytic materials.

2.6.1 Applications to Nanoparticle Imaging

Functional polymeric nanoparticles (Figure 2.11a) are widely employed for a variety of biomedical applications such as targeted drug delivery. However, the analysis of the structure and functionality of such nanoparticles is particularly challenging due to the small size and complexity of the structure. Localization microscopy inspired by DNA-PAINT and qPAINT enables the direct visualization and counting of functional groups at the single-particle level, as well as resolving size and morphology. In a recent study it was shown that the amount of internalized compound and the number of functional groups on the surface of PLGA-PEG nanoparticles was strongly heterogeneous, with some particles not incorporating any compound at all [44]. Similar studies were reported for cetuximab functionalized particles targeting cancer-related cell receptors [45] and metallic nanoparticles functionalized with DNA [42, 46].

After successful functionalization and drug internalization, the nanoparticles are often injected in the blood stream where they accumulate a

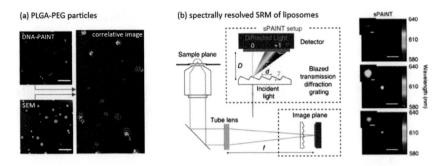

FIGURE 2.11 (a) correlative DNA-PAINT and scanning electron microscopy of PLGA-PEG nanoparticles. (b) Spectrally resolved PAINT (sPAINT) of liposomes. ([a] Reproduced with permission under a Creative Commons CC-BY-NC-ND License from [44]. Copyright The Authors; [b] Adapted with permission under a Creative Commons CC-BY License from [48]. Copyright The Authors.)

so-called protein corona. The density and composition of the protein corona was recently also studied using localization microscopy, where again a large degree of heterogeneity in the amount and type of protein per particle was identified [47]. These studies will aid in further optimization of the drug carrier itself, the functional targeting groups, the antifouling coating, and their interaction with a complex matrix (like blood) at the single-particle and single-molecule level.

Due to their biocompatibility liposomes and lipid nanoparticles (Figure 2.11b) also play a crucial role in drug delivery applications. The typical liposome size range (50–200 nm) can be addressed perfectly by SRM. The hydrophobic nature of the bilayer interior of liposomes makes these assemblies ideally suited for PAINT using lipophilic probes. The first examples of PAINT involved the use of hydrophobic probe Nile Red to image 100-nm-diameter large unilamellar vesicles and lipid bilayers [48]. Nile Red is an ideal probe in PAINT, because its fluorescence emission is almost negligible in water but strong in apolar environments. Moreover, the Nile Red association time to lipids of approximately 10 ms is suitable for on/off switching, thus enabling single-molecule localization.

In addition, the emission of Nile-red depends strongly on the hydrophobicity of its local environment. For this reason, spectrally resolved PAINT (sPAINT) has been introduced by placing a diffraction grating right before the camera. The insertion of a transmission diffraction grating into the optical path (Figure 2.11b) enables the acquisition of fluorescence spectra of individual solvatochromic fluorophores and leads to high-resolution maps of the hydrophobicity of large unilamellar vesicles and other structures, such as β-amyloid(1–42) fibers, α-synuclein fibrils, and the cell plasma membrane [49]. This pioneering example highlights how SRM can not only reveal structural information but also probe the chemical properties of nanomaterials by exploiting single-molecule spectroscopy. A more in-depth description on the use of SRM for nanomedicine can be found in Chapter 11.

2.6.2 Applications to Polymer Imaging

The characterization of polymeric materials (Figure 2.12) is key toward the understanding of structure–activity relations and therefore to the rational design of novel and improved materials for a myriad of applications. For example, the charge transport in conjugated polymers was studied using localization microscopy [50]. Single carrier dynamics was studied with high spatial and temporal resolution, while providing structural information on single conjugated polymer chains. The method first tracks nanoscale hole polaron motion at 1 kHz framerate and then utilizes binding and unbinding dynamics of water-soluble quenchers onto the nanostructure to reconstruct a superresolution map of emission sites throughout the structure. By overlaying the hole polaron trajectories with the superresolution maps, structural information can be correlated with charge transport properties, as shown in Figure 2.12a.

Supramolecular polymers on the other hand are polymers in which the monomers are linked by non-covalent bonds, typically by hydrophobic interactions and hydrogen bonding. Their dynamic nature, together with their modularity and responsiveness to different stimuli, makes them promising

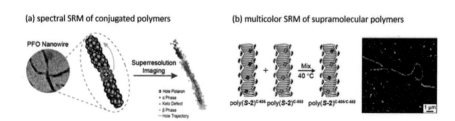

FIGURE 2.12 (a) spectrally resolved STORM imaging to study charge transport in conjugated polymers. (b) Multi-color localization microscopy to study the dynamics of monomer exchange in supramolecular polymers immersed in organic solvents. ([a] Adapted with permission from [50]. Copyright 2021 American Chemical Society; [b] Reprinted with permission under a Creative Commons CC-BY-NC-ND License from [52]. Copyright 2018 American Chemical Society.)

candidates for several applications in biomedicine, optoelectronics, sensing, and catalysis. In 2014 the first example of supramolecular polymer imaging using SRM was demonstrated [51], where the monomer exchange mechanism of water-soluble 1,3,5-benzenetricarboxamide supramolecular polymers were studied using 2-color dSTORM. The combination of dSTORM and stochastic modeling demonstrated that a homogeneous exchange occurs all along the BTA self-assembled fibrillar structures. Follow-up studies unveiled how the dynamics of supramolecular polymers is influenced by organic solvents (Figure 2.12b) [52], hydrophobicity, chirality, the presence of functional groups, and the chemical structure of the monomer. In all cases, the mechanism and the kinetics of monomer exchange were unveiled using multi-color localization microscopy, and the structure–dynamics relationships were highlighted with unprecedented spatial resolution.

Although polymers are usually fluorescently labeled in order to be studied with SRM, some polymers may exhibit intrinsic fluorescence in the visible range. PMMA, polystyrene and Su-8 films were imaged using STORM without any external labeling, showing intensive blinking events that enabled the reconstruction of a well-defined image. Therefore, the development of novel blinking dyes that are emissive in the polymeric environment and the development of self-blinking materials suitable for label-free SMLM represent two interesting perspectives in SRM. STORM has also been employed to study. A more detailed description of the use of localization microscopy for supramolecular and polymer science can be found in Chapter 10.

REFERENCES

1. P. Sengupta, S. B. Van Engelenburg, and J. Lippincott-Schwartz, "Superresolution imaging of biological systems using photoactivated localization microscopy," *Chem. Rev.* **114**, 3189–3202 (2014).
2. B. E. A. Saleh and M. C. Teich, *Fundamentals of Photonics*, Wiley Series in Pure and Applied Optics (John Wiley & Sons, Inc., 1991).
3. S. Stallinga and B. Rieger, "Accuracy of the Gaussian point spread function model in 2D localization microscopy," *Opt. Express* **18**, 24461 (2010).
4. K. I. Mortensen, L. S. Churchman, J. A. Spudich, and H. Flyvbjerg, "Optimized localization analysis for single-molecule tracking and super-resolution microscopy," *Nat. Methods* **7**, 377–381 (2010).
5. A. Small and S. Stahlheber, "Fluorophore localization algorithms for super-resolution microscopy," *Nat. Methods* **11**, 267–279 (2014).
6. R. J. Ober, S. Ram, and E. S. Ward, "Localization accuracy in single-molecule microscopy," *Biophys. J.* **86**, 1185–1200 (2004).

7. K. A. Winick, "Cramér–Rao lower bounds on the performance of charge-coupled-device optical position estimators," *J. Opt. Soc. Am. A* **3**, 1809 (1986).
8. H. Cramér, *Mathematical Methods of Statistics*, Princeton Mathematical Series No. 9 (Princeton University Press, 1946).
9. C. Radhakrishna Rao, "Information and accuracy attainable in the estimation of statistical parameters," *Bull. Calcutta Math. Soc.* **37**, 81–91 (1945).
10. A. V. Abraham, S. Ram, J. Chao, E. S. Ward, and R. J. Ober, "Quantitative study of single molecule location estimation techniques," *Opt. Express* **17**, 23352 (2009).
11. C. S. Smith, N. Joseph, B. Rieger, and K. A. Lidke, "Fast, single-molecule localization that achieves theoretically minimum uncertainty," *Nat. Methods* **7**, 373–375 (2010).
12. H. Deschout, F. Cella Zanacchi, M. Mlodzianoski, A. Diaspro, J. Bewersdorf, S. T. Hess, and K. Braeckmans, "Precisely and accurately localizing single emitters in fluorescence microscopy," *Nat. Methods* **11**, 253–266 (2014).
13. M. Lelek, M. T. Gyparaki, G. Beliu, F. Schueder, J. Griffié, S. Manley, R. Jungmann, M. Sauer, M. Lakadamyali, and C. Zimmer, "Single-molecule localization microscopy," *Nat. Rev. Methods Primer* **1**, 39 (2021).
14. R. E. Thompson, D. R. Larson, and W. W. Webb, "Precise nanometer localization analysis for individual fluorescent probes," *Biophys. J.* **82**, 2775–2783 (2002).
15. S. M. Früh, U. Matti, P. R. Spycher, M. Rubini, S. Lickert, T. Schlichthaerle, R. Jungmann, V. Vogel, J. Ries, and I. Schoen, "Site-specifically-labeled antibodies for super-resolution microscopy reveal *in situ* linkage errors," *ACS Nano* **15**, 12161–12170 (2021).
16. M. Shang, Z. Huang, and Y. Wang, "Influence of drift correction precision on super-resolution localization microscopy," *Appl. Opt.* **61**, 3516 (2022).
17. H. Deschout, K. Neyts, and K. Braeckmans, "The influence of movement on the localization precision of sub-resolution particles in fluorescence microscopy," *J. Biophotonics* **5**, 97–109 (2012).
18. W. Vandenberg, M. Leutenegger, T. Lasser, J. Hofkens, and P. Dedecker, "Diffraction-unlimited imaging: from pretty pictures to hard numbers," *Cell Tissue Res.* **360**, 151–178 (2015).
19. E. Betzig, G. H. Patterson, R. Sougrat, O. W. Lindwasser, S. Olenych, J. S. Bonifacino, M. W. Davidson, J. Lippincott-Schwartz, and H. F. Hess, "Imaging intracellular fluorescent proteins at nanometer resolution," *Science* **313**, 1642–1645 (2006).
20. M. J. Rust, M. Bates, and X. Zhuang, "Sub-diffraction-limit imaging by stochastic optical reconstruction microscopy (STORM)," *Nat. Methods* **3**, 793–796 (2006).
21. M. Bates, T. R. Blosser, and X. Zhuang, "Short-range spectroscopic ruler based on a single-molecule optical switch," *Phys. Rev. Lett.* **94**, 108101 (2005).
22. M. Heilemann, E. Margeat, R. Kasper, M. Sauer, and P. Tinnefeld, "Carbocyanine dyes as efficient reversible single-molecule optical switch," *J. Am. Chem. Soc.* **127**, 3801–3806 (2005).
23. M. Heilemann, S. van de Linde, M. Schüttpelz, R. Kasper, B. Seefeldt, A. Mukherjee, P. Tinnefeld, and M. Sauer, "Subdiffraction-resolution

fluorescence imaging with conventional fluorescent probes," *Angew. Chem. Int. Ed.* **47**, 6172–6176 (2008).

24. G. T. Dempsey, M. Bates, W. E. Kowtoniuk, D. R. Liu, R. Y. Tsien, and X. Zhuang, "Photoswitching mechanism of cyanine dyes," *J. Am. Chem. Soc.* **131**, 18192–18193 (2009).

25. J. Schnitzbauer, M. T. Strauss, T. Schlichthaerle, F. Schueder, and R. Jungmann, "Super-resolution microscopy with DNA-PAINT," *Nat. Protoc.* **12**, 1198–1228 (2017).

26. F. Schueder, J. Stein, F. Stehr, A. Auer, B. Sperl, M. T. Strauss, P. Schwille, and R. Jungmann, "An order of magnitude faster DNA-PAINT imaging by optimized sequence design and buffer conditions," *Nat. Methods* **16**, 1101–1104 (2019).

27. S. Coelho, J. Baek, J. Walsh, J. J. Gooding, and K. Gaus, "3D active stabilization for single-molecule imaging," *Nat. Protoc.* **16**, 497–515 (2021).

28. C. Kuo and R. M. Hochstrasser, "Super-resolution microscopy of lipid bilayer phases," *J. Am. Chem. Soc.* **133**, 4664–4667 (2011).

29. A. S. Eklund, M. Ganji, G. Gavins, O. Seitz, and R. Jungmann, "Peptide-PAINT super-resolution imaging using transient coiled coil interactions," *Nano Lett.* **20**, 6732–6737 (2020).

30. R. Riera, T. P. Hogervorst, W. Doelman, Y. Ni, S. Pujals, E. Bolli, J. D. C. Codée, S. I. Van Kasteren, and L. Albertazzi, "Single-molecule imaging of glycan–lectin interactions on cells with Glyco-PAINT," *Nat. Chem. Biol.* **17**, 1281–1288 (2021).

31. G. C. Rollins, J. Y. Shin, C. Bustamante, and S. Pressé, "Stochastic approach to the molecular counting problem in superresolution microscopy," *Proc. Natl. Acad. Sci.* **112**, (2015).

32. R. Jungmann, M. S. Avendaño, M. Dai, J. B. Woehrstein, S. S. Agasti, Z. Feiger, A. Rodal, and P. Yin, "Quantitative super-resolution imaging with qPAINT," *Nat. Methods* **13**, 439–442 (2016).

33. U. B. Choi, J. J. McCann, K. R. Weninger, and M. E. Bowen, "Beyond the random coil: Stochastic conformational switching in intrinsically disordered proteins," *Structure* **19**, 566–576 (2011).

34. P. R. Nicovich, D. M. Owen, and K. Gaus, "Turning single-molecule localization microscopy into a quantitative bioanalytical tool," *Nat. Protoc.* **12**, 453–460 (2017).

35. Y.-L. Wu, A. Tschanz, L. Krupnik, and J. Ries, "Quantitative data analysis in single-molecule localization microscopy," *Trends Cell Biol.* **30**, 837–851 (2020).

36. R. Riera, E. Archontakis, G. Cremers, T. De Greef, P. Zijlstra, and L. Albertazzi, "Precision and accuracy of receptor quantification on synthetic and biological surfaces using DNA-PAINT," *ACS Sens.* **8**, 80–93 (2023).

37. M. Zhang, M. Lihter, T.-H. Chen, M. Macha, A. Rayabharam, K. Banjac, Y. Zhao, Z. Wang, J. Zhang, J. Comtet, N. R. Aluru, M. Lingenfelder, A. Kis, and A. Radenovic, "Super-resolved optical mapping of reactive sulfur-vacancies in two-dimensional transition metal dichalcogenides," *ACS Nano* **15**, 7168–7178 (2021).

38. A. M. Arnold, M. C. Schneider, C. Hüsson, R. Sablatnig, M. Brameshuber, F. Baumgart, and G. J. Schütz, "Verifying molecular clusters by 2-color localization microscopy and significance testing," *Sci. Rep.* **10**, 4230 (2020).

39. S.-H. Lee, J. Y. Shin, A. Lee, and C. Bustamante, "Counting single photo-activatable fluorescent molecules by photoactivated localization microscopy (PALM)," *Proc. Natl. Acad. Sci. U. S. A.* **109**, 17436–17441 (2012).

40. L. G. Jensen, T. Y. Hoh, D. J. Williamson, J. Griffié, D. Sage, P. Rubin-Delanchy, and D. M. Owen, "Correction of multiple-blinking artifacts in photoacti-vated localization microscopy," *Nat. Methods* **19**, 594–602 (2022).

41. P. Sengupta, T. Jovanovic-Talisman, D. Skoko, M. Renz, S. L. Veatch, and J. Lippincott-Schwartz, "Probing protein heterogeneity in the plasma membrane using PALM and pair correlation analysis," *Nat. Methods* **8**, 969–975 (2011).

42. M. Horáček, D. J. Engels, and P. Zijlstra, "Dynamic single-molecule count-ing for the quantification and optimization of nanoparticle functionalization protocols," *Nanoscale* **12**, 4128–4136 (2020).

43. K. Xu, G. Zhong, and X. Zhuang, "Actin, Spectrin, and associated proteins form a periodic cytoskeletal structure in axons," *Science* **339**, 452–456 (2013).

44. T. Andrian, P. Delcanale, S. Pujals, and L. Albertazzi, "Correlating super-resolution microscopy and transmission electron microscopy reveals multi-parametric heterogeneity in nanoparticles," *Nano Lett.* **21**, 5360–5368 (2021).

45. L. Woythe, P. Madhikar, N. Feiner-Gracia, C. Storm, and L. Albertazzi, "A single-molecule view at nanoparticle targeting selectivity: correlating ligand functionality and cell receptor density," *ACS Nano* **16**, 3785–3796 (2022).

46. A. Taylor, R. Verhoef, M. Beuwer, Y. Wang, and P. Zijlstra, "All-optical imaging of gold nanoparticle geometry using super-resolution microscopy," *J. Phys. Chem. C* **122**, 2336–2342 (2018).

47. N. Feiner-Gracia, M. Beck, S. Pujals, S. Tosi, T. Mandal, C. Buske, M. Linden, and L. Albertazzi, "Super-resolution microscopy unveils dynamic heteroge-neities in nanoparticle protein corona," *Small* **13**, 1701631 (2017).

48. M. N. Bongiovanni, J. Godet, M. H. Horrocks, L. Tosatto, A. R. Carr, D. C. Wirthensohn, R. T. Ranasinghe, J. E. Lee, A. Ponjavic, J. V. Fritz, C. M. Dobson, D. Klenerman, and S. F. Lee, "Multi-dimensional super-resolution imaging enables surface hydrophobicity mapping," *Nat. Commun.* **7**, 1–9 (2016).

49. S. Moon, R. Yan, S. J. Kenny, Y. Shyu, L. Xiang, W. Li, and K. Xu, "Spectrally resolved, functional super-resolution microscopy reveals nanoscale compo-sitional heterogeneity in live-cell membranes," *J. Am. Chem. Soc.* **139**, 10944–10947 (2017).

50. Y. Jiang, H. Chen, X. Men, Z. Sun, Z. Yuan, X. Zhang, D. T. Chiu, C. Wu, and J. McNeill, "Multimode time-resolved superresolution microscopy revealing chain packing and anisotropic single carrier transport in conjugated poly-mer nanowires," *Nano Lett.* **21**, 4255–4261 (2021).

51. L. Albertazzi, R. W. V. D. Hofstad, and E. W. Meijer, "probing exchange path-ways in super-resolution microscopy," *Science* **491**, 10–15 (2014).

52. B. Adelizzi, A. Aloi, N. J. Van Zee, A. R. A. Palmans, E. W. Meijer, and I. K. Voets, "Painting supramolecular polymers in organic solvents by super-resolution microscopy," *ACS Nano* **12**, 4431–4439 (2018).

CHAPTER 3

Stimulated Emission Depletion Microscopy

Lorenzo Albertazzi and Peter Zijlstra

Eindhoven University of Technology, Eindhoven, The Netherlands

3.1 INTRODUCTION

In Chapter 2 we discussed localization microscopy approaches that switch the majority of fluorophores in the field of view to a dark state using optical or chemical switching mechanisms. A small fraction of fluorophores enters the bright state stochastically, enabling the reconstruction of a high-resolution image from many diffraction limited images. Stimulated emission depletion nanoscopy (STED) is also based on controlling the on/off state of fluorophores, but in a deterministic rather than stochastic way. This approach to circumvent the diffraction limit was proposed in 1994 [1] and first realized in 1999 [2, 3]; it uses stimulated emission to reversibly silence fluorophores in predefined regions in the sample to facilitate separation at sub-diffraction length scales.

In this chapter we will review the basics of STED microscopy, starting with the underlying principles. We then discuss the optical hardware needed and describe the resolution that can be achieved. We discuss the different mechanisms that can be used to silence fluorophores, each with their own advantages and disadvantages. We describe new developments in terms of variations on the STED principle, and finish with select examples on the use of STED in the fields of materials science and physical chemistry.

DOI: 10.1201/9781003220688-3

3.2 PRINCIPLES OF STED

3.2.1 Stimulated Emission Depletion

In a regular confocal microscope, the excitation volume is diffraction limited, with a typical dimension given by Abbe's diffraction limit ($r = \lambda/2\mathrm{NA}$). This diffraction limited excitation spot (Figure 3.1a) excites fluorophores whose Stokes shifted emission is detected on a photodiode. By scanning the focal spot with respect to the sample, an image is created with a resolution equal to Abbe's diffraction limit. In a typical implementation of STED nanoscopy, a second laser (the STED laser) is added to the microscope, which forces excited fluorophores to their dark ground state (i.e., it minimizes their fluorescence).

The approximate shape of the used laser spots is shown in Figure 3.1b and c. The excitation spot is a regular (nearly) Gaussian spot that results from the diffraction of a plane wave from the circular aperture of the objective lens. The focal spot of the STED laser is typically shaped like a doughnut (we will discuss the means to create such a spot in Section 3.2.2), which results in depletion in the perimeter of the excitation spot. When the excitation and depletion beams are overlapped in three-dimensional space, this results in fluorescence emission from a volume that is significantly smaller than the diffraction-limited excitation volume. The size of this effective spot is typically 20–50 nm but depends on the fluorophores that are used and on the depletion intensity and wavelength (see Section 3.2.3). Note that both the excitation beam and the STED beam are focused by a regular objective lens and are therefore diffraction limited.

The mechanism by which the fluorescence is minimized is based on stimulated emission, as depicted in the Jablonski diagram in Figure 3.1d. The excitation beam (in blue) excites the fluorophore from the ground state to a higher vibrational level in the first excited state. After fast internal conversion, the excited state decays back to the ground state on timescales of approximately 1 nanosecond, yielding a fluorescence photon. Depletion can be achieved by a high-power STED beam that stimulates the excited fluorophore back to the ground state. Stimulated decay back to the ground state then results in the emission of a photon that has the same energy, polarization, and wavevector as the photon that stimulated the emission. As shown in Figure 3.1e, the STED laser is often tuned to the red wing of the emission spectrum of the fluorophore. This enables the use of high-quality optical filters to prevent the STED laser beam and the stimulated emission photons from reaching the detector.

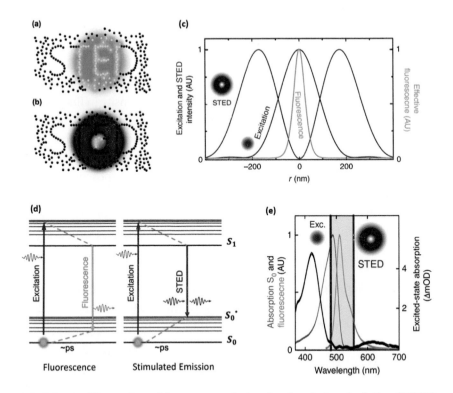

FIGURE 3.1   Illustration of the concept of stimulated emission depletion (STED) nanoscopy. (a) In a regular confocal microscope all fluorophores that reside in the excitation volume are excited and fluoresce. This results in a diffraction-limited probe volume indicated by the green area. (b) In a STED microscope the excitation volume is the same, but all fluorophores in the perimeter are depleted by stimulated emission that is induced by the red doughnut beam. (c) Cross section of the excitation beam and the doughnut-shaped STED beam. Because the doughnut-shaped STED beam de-excites the fluorophores in high-intensity regions, fluorescence photons are only emitted from the center of the doughnut beam (where the STED intensity is low) resulting in detectable fluorescence from a spot that is effectively smaller than the diffraction limit. (d) Jablonski diagram indicating the energy of the excitation and STED beams, as well as the fluorescence. (e) Corresponding spectra. The graph shows the wavelengths of the excitation and STED beams together with the absorption and fluorescence spectra of a typical dye. The STED laser is tuned to the red wing of the emission spectrum where it induces stimulated emission. Because excited-state absorption can cause rapid bleaching of the dye, it is important that the excitation and STED wavelengths are far from the excited-state absorption. ([a], [b], and [d] Adapted with permission under a Creative Commons CC-BY license from [4], copyright The Authors. [c] and [e] Adapted with permission from [5]. Copyright 2018 Springer Nature.)

Efficient stimulated emission depletion requires a high-power STED beam to increase the probability that stimulated emission is induced during the very short nanosecond lifetime of the excited state of the fluorophore. Typical intensities of the STED laser can be > 100 MW/cm^2 (similar to laser-based marble cutters!), whereas the excitation beam intensity in fluorescence microscopy is typically <0.01 MW/cm^2. Therefore, a concern has been that the STED laser might introduce phototoxic effects, such as photobleaching of the fluorescent labels, production of radicals, light-induced cell death, sample heating or optical trapping. Consequently, STED nanoscopy was long thought to be incompatible with fragile samples such as polymers and live cells. Later we will discuss the mechanism that enabled a strong reduction in the required STED intensity facilitating the imaging of fragile samples.

3.2.2 Optical Setup

A STED microscope is constructed around a conventional confocal microscope as described in Chapter 1. The excitation and the STED laser beams are overlapped using two dichroic mirrors and directed to the objective lens. In the sample plane the spatial overlap of the excitation and STED laser spots results in a reduced effective focal spot. The objective lens then captures the fluorescence, the reflected STED beam and a fraction of the stimulated emission photons, which are separated by the dichroic mirrors and (if needed) additional filters before reaching the detector. This ensures that only the fluorescence photons from the very small effective focal volume reach the detector.

The doughnut-shaped STED beam is most often created by a vortex phase plate in the STED-laser beam path. This phase plate alters the phase of the STED beam in a location-dependent manner, with the result that the focused light interferes destructively in the focus center. The phase differences introduced by such a vortex phase plate are illustrated in Figure 3.2b, together with the resulting intensity profile around the focal plane. The intensity profile in the xy-plane is doughnut-shaped, but from the profile in the xz-plane (along the optical axis) indicates that this phase plate can only provide depletion in the xy-plane because the intensity above and below the focal plane are low. This phase plate therefore enables STED in two dimensions, whereas 3D-STED will require alternative solutions discussed in Section 3.2.4.

The main difficulty in aligning a STED microscope is the required nanometer-scale spatial overlap between the excitation and STED laser beams. Specialty phase plates could make this significantly easier by employing

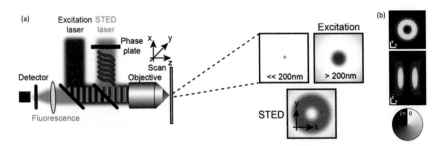

FIGURE 3.2 (a) Simplified schematic of the optical setup used for STED micros-copy. The excitation and STED laser beams are combined using dichroic mir-rors. An image is formed by scanning the focused laser beams over the sample, or by scanning the sample through the focused laser beams. The fluorescence is detected by a point detector, often a photon counter. The insets show the effec-tive spot from which the fluorescence is detected (in green), the near-Gaussian shaped excitation beam (in blue) and the doughnut-shaped STED beam (in orange). (b) The doughnut shaped STED beam is generated by a spiral phase plate that introduces a phase delay on the incident beam that varies from zero to 2π. This results in a doughnut shaped focus in the x-y plane. (Adapted with permission under a Creative Commons License CC-NC-ND from [8]. Copyright Christian Eggeling et al.)

materials with a strong wavelength-dependent response (dispersion) that modulates the phase of the STED beam while leaving the bluer excitation beam unaffected [6]. Alternatively, a polarization-dependent response (bire-fringence) can be exploited to modulate orthogonally polarized excitation and STED laser beams [7]. In both cases the laser beams can be combined first in, for example, an optical fiber, and the phase plate is introduced directly in front of the objective lens.

Different excitation schemes have been developed over the years employ-ing continuous wave, pulsed, and gated excitation schemes (Figure 3.3). Since the signal is optimized when depletion occurs shortly after fluoro-phore excitation, STED microscopy is typically implemented with syn-chronized and temporally aligned pulsed lasers (p-STED in Figure 3.3) where the excitation pulses are followed immediately by the depletion pulses. In the beginning of the 2000s, most STED microscopes used pulsed and tunable titanium sapphire mode-locked lasers as STED beams, whose pulses required stretching to guarantee a few hundred picosecond pulse width and conversion to the visible range if imaging with green–yellow fluorophores [3].

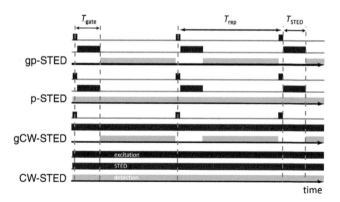

FIGURE 3.3   Examples of possible excitation schemes. The bottom rows show the traditional continuous-wave STED (CW-STED), in which both laser beams are on at all times, and fluorescence is detected at all times. Second from the bottom shows gated CW-STED (gCW-STED), where the excitation beam is pulsed, and fluorescence is detected after a certain time T_{gate}. The third row from the bottom displays pulsed STED (p-STED) in which both the excitation and STED lasers are pulsed, whereas fluorescence is detected continuously. Finally, the top row shows gated pulsed STED (gp-STED) combining pulsed laser excitation with gated detection. (Adapted with permission from [5]. Copyright 2018 Springer Nature.)

A huge simplification to STED nanoscopy has been introduced by realizing it with continuous-wave (CW) lasers, making laser pulse preparation redundant [9]. CW-STED (Figure 3.3) has now been demonstrated with compact fibre lasers or diode-pumped solid-state lasers. However, CW-STED nanoscopy lags somewhat behind its pulsed STED counterpart in performance due to a lower probability of stimulated emission induced by a CW-laser compared to a pulsed laser. In p-STED, the synchronized excitation and STED pulses realize an instantaneous and thus optimized efficiency of fluorescence switching, since the pulses of the STED beam reach the focal plane a few picoseconds after the excitation pulses so as to instantly deplete the excited state. Consequently, approximately three- to five-fold higher average laser powers have to be applied for CW-STED to ensure efficient depletion.

Furthermore, in the CW-STED modality, a non-negligible part of the molecules emits fluorescence before having been exposed to much of the STED light, and thus residual fluorescence outside the zero-intensity point of the STED beam reduces the resolution. A remedy to these limitations is the use of a CW-STED laser in combination with a pulsed excitation laser and a gated detection scheme (gated CW-STED or gCW-STED in Figure 3.3).

This still avoids careful synchronization of laser pulses and allows the use of compact and less costly STED laser systems. Roughly speaking, by removing the fluorescence occurring during the STED beam action, time gating increases the depletion without increasing the STED beam intensity. Such gating applies to both pulsed and CW laser excitation schemes, but the latter represents the cheapest and simplest implementation so far.

3.2.3 Optical Resolution

The resolution of a STED image is no longer governed by Abbe's diffraction limit but is logically smaller than that. For STED microscopy this apparent resolution depends on the STED beam intensity that depletes the fluorophores in the doughnut beam: the higher the STED power the more molecules are depleted and the higher the resolution. Therefore, STED resolution can be tuned to obtain the best tradeoff between resolution and illumination power.

This is shown in Figure 3.4a–g, where a standard confocal image of 24 nm fluorescent beads is compared to STED images recorded with different STED laser intensities. It can be clearly seen that the ability to separate two beads increases with increasing STED power, in other words the resolution increases. The diameter of the bead image is plotted as a function of STED laser power in Figure 3.4h, where a clear improvement in

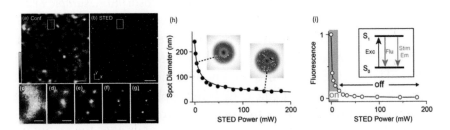

FIGURE 3.4 (a–b) Comparison between regular confocal and STED imaging of 24 nm fluorescent beads on a coverslip. (c–g) STED images of the same region of the sample recorded for increasing STED intensities. The resolution gain can directly be observed. (h–i) Fluorescence intensity and spot diameter as a function of STED power. Higher STED powers result in a smaller spot diameter but simultaneously in a lower fluorescence intensity because most fluorophores in the excitation volume get depleted. Scale bar in (a, b) 1 μm, in (c–g) 200 nm. ([a–g] Reproduced with permission from [11]. Copyright 2008 Optical Society of America; [h] and [i] Adapted with permission under a Creative Commons License CC-NC-ND from [8]. Copyright Christian Eggeling et al.)

resolution can be observed. For increasing STED laser power, the full width at half maximum of the fluorescent spot can be described by

$$d \approx \frac{\lambda}{2\text{NA}} \frac{1}{\sqrt{I_{\text{STED}}^{\max}/I_{\text{sat}}+1}}, \tag{3.1}$$

where the reduction in resolution is determined by the saturation factor $I_{\text{STED}}^{\max}/I_{\text{sat}}$, which is the ratio between the maximum intensity in the STED depletion beam ($I_{\text{STED}}^{\max}$) and the intensity at which the probability of fluorescence emission is reduced by half (I_{sat}). The latter is an intrinsic feature of fluorophores and describes the tendency of a fluorophore to undergo stimulated emission. Note that, in the limit of zero STED intensity, the above equation reduces to the familiar diffraction limit. As shown in Figure 3.4h the measured spot size is in good agreement with the predictions from Eq. 3.1. The derivation of Eq. 3.1 assumes a $\sin^2 x$ profile of the STED beam (with a zero intensity at the origin) and a $\cos^2 x$ profile of the excitation beam (with a maximum at the origin) near the focus. Assuming that the fluorescence probability reduces exponentially with STED intensity then yields Eq. 3.1 as shown in [10].

In theory, the maximum obtainable resolution as stated by Eq. 3.1 would be infinitely small. In practice however, the extreme intensities required to reach high resolutions would lead to unwanted effects such as photo-bleaching, optical trapping, multi-photon absorption, sample heating, or even sample destruction. The chosen fluorophore will also largely determine the maximal obtainable resolution because some fluorophores are more sensitive to the STED depletion lasers than others. Regardless of the STED intensity: there will always be a fraction of fluorophores in the depletion area that are not efficiently being depleted. Additionally, some fluorophores in the depletion area may be re-excited (anti-Stokes) by the depletion light. If the emission filters are non-perfect, then some stimulated light will still reach the detectors. In practice, the maximum obtainable resolution is therefore 20–40 nm.

In addition to an improved resolution, the depletion of fluorophores by the STED beam also has consequences for the detected fluorescence intensity, see Figure 3.4i. Increasing STED power depletes fluorophores in the doughnut beam that do not contribute to the fluorescence signal anymore. As a result, improving the resolution always comes at the cost of signal intensity. For this reason, STED images are often cleaned up using deconvolution to improve the final signal-to-noise ratio.

3.2.4 3D Imaging

So far, we have considered 2D imaging with STED beam that is dough-nut shaped in the sample plane only. As can be seen in Figure 3.2b, this STED beam configuration has near-zero intensity above and below the focus, and thus does not result in increased resolution along the optical axis of the microscope. In order to expand STED microscopy to the 3rd dimension, a different STED beam shape is required that forms a 'zero'-intensity point surrounded in all directions by regions with high intensity. Currently, two approaches have been followed. The first uses two superimposed incoherent STED beams, one producing the doughnut-shaped focus using a vortex phase plate to confine the fluorescence laterally, whereas the other beam produces a bottle-shaped focus and confines the fluorescence axially (Figure 3.5a, b). This approach is currently available in different commercial systems.

The resulting 3D images are shown in Figure 3.5c for 20 nm fluorescent spheres on glass. Images in confocal (with the STED beams disabled) and STED are shown in the xy-and xz-plane. In addition to an improvement in the lateral resolution, a significant reduction in cross-sectional area in the xz-plane is also observed due to the superposition of the two STED beams. The right graph Figure 3.5d shows the focal volume reduction relative to confocal microscopy, where the combination of two STED beams at a 70:30 intensity ratio gives a maximal volume reduction factor of 125.

For 3D applications that include thick samples such as polymeric films and hydrogels, the resolution improvement must be preserved deep into the sample. Similar to conventional microscopy, light scattering and aberrations are the limiting factors for deep STED microscopy imaging. In addition, scattering and aberrations can break the coalignment of the excitation and STED beams and degrade the quality of the 'zero'-intensity point. The first approach for reducing sample-induced spherical aberration (due to refractive index mismatch between the sample and objective lens immersion medium) is the use of a manual correction collar in the objective lens [12]. This approach allows reaching a sub-diffraction resolution in thick samples, but it has been shown to be effective only for 2D STED where the doughnut-shaped STED beam is less sensitive to aberration compared to the bottle-shaped STED beam. Furthermore, in the case of large-scale 3D images, manual correction at each axial position is unrealistic.

A more general solution, valid both for 2D- and 3D-STED imaging and also able to correct system-induced aberrations, is the use of adaptive optics based on deformable mirrors or a spatial light modulator.

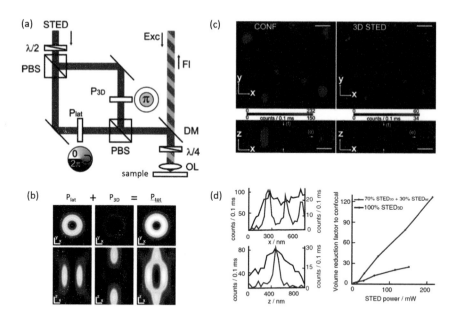

FIGURE 3.5   (a) Optical setup used for 3D STED, in which two STED beams with a different phase mask in the beam path are overlapped to generate a 3D doughnut beam in the focus. The two beams are combined on a polarizing beam splitter (PBS) and reflected to the objective lens (OL) by a dichroic mirror (DM). The fluorescence (FL) is detected in reflection. (b) Calculated intensity distributions in the xy- and xz-plane of the two different STED beams. As shown in (a), the phase mask for 2D-STED resulting in the power density P_{lat} is a helical retardation from 0 to 2π, whereas for 3D-STED an additional beam shaped by a phase mask with two concentric circles is added. The total power density P_{tot} then results in a 3D doughnut beam. Calculations were done for a STED wavelength of 775 nm and an NA of 1.4. The look-up tables are identical for all images. Scale bars are 200 nm. (c) 3D nanoscale image of 20 nm fluorescent spheres on glass. Images in confocal (with the STED beams disabled) and STED are shown in the xy- and xz-plane. Note the significant reduction in cross-sectional area in the STED xz-image. Scale bars 1 μm. (d) Intensity profile along the x- and z-direction. The right graph shows the focal volume reduction relative to confocal microscopy, where the combination of two STED beams gives a maximal volume reduction factor of 125. (Panels (a), (c) and (d) Reprinted with permission from [14]. Copyright 2008 American Chemical Society. Panel (b) reproduced with permission from [15]. Copyright 2018 AIP Publishing.)

These solutions introduce a location-dependent phase modulation, very similar to the phase plates shown in Figure 3.5a but in a programmable manner [13]. This allows for the dynamic finetuning of the phase modulation on the STED beam and enables the correction of aberrations in nearly any environment.

3.2.5 Other State Transitions

In the most prominent implementations of STED microscopy, stimulated emission is induced to deplete the fluorescence from the fluorophores residing in the STED doughnut beam. This is a generic principle because it only requires fluorophores that have a ground and first excited state. These are the basic states present in all dyes, but the general applicability comes at a cost that was mentioned before: the short (nanosecond) lifetime of the first excited state requires a very high STED intensity to achieve efficient depletion. Once it is clear that this is a general principle, it is obvious that stimulated emission is not the only way by which we can achieve depletion. There are other "on" and "off" states in a dye which one can use to the same effect, notably triplet states and cis-trans conformational states, as shown in Figure 3.6.

Ground-state depletion microscopy employs metastable triplet states (see the Jablonski diagram in Figure 3.6b) with a lifetime of microseconds to deplete the excited state of the fluorophore. The result is that the saturation intensity I_{sat} in Eq. 3.1 reduces from MW/cm^2 to kW/cm^2. This makes the use of transitions between long-lived states very attractive because it overcomes one of the main drawbacks of STED microscopy, the extremely high-power densities required that may induce fluorophore and sample damage.

A further reduction in the required depletion intensity can be obtained using cis-trans isomerization in switchable fluorescent proteins. These can

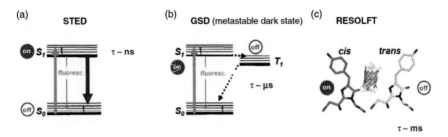

FIGURE 3.6 States and state transitions utilized in (a) STED, (b) ground-state depletion (GSD) and (c) Reversible Saturable/Switchable Optically Linear (Fluorescence) Transitions (RESOLFT) nanoscopy. The intensity required to switch the fluorophore from the bright to the depleted form is inversely related to the state lifetime. The longer the lifetime of the involved states, the lower intensity needed to establish the switch required to deplete the fluorescence from regions within the doughnut beam. (Reproduced with permission under a Creative Commons License CC-BY from [16]. Copyright 2019 The Authors.)

have lifetimes of milliseconds, further reducing the required depletion intensity to W/cm². This equates to a reduction of six orders of magnitude compared to the use of stimulated emission from the first excited to the ground state. The microscopy implementations exploiting cis-trans isomerization are referred to as RESOLFT microscopy, which stands for Reversible Saturable/Switchable Optically Linear Fluorescence Transitions.

This not only enables the use of low-power continuous-wave laser beams for nanoscopy, but also enables the parallel scanning of many beams simultaneous. The low power requirement allows one to split the depletion laser into multiple beams and scan them in parallel over different areas on the sample. As long as the doughnuts are separated by more than Abbe's diffraction limit, they can be read out simultaneously by projecting the signal generated in this array of minima onto a camera. Only a limited number of scanning steps is required to create an image, a recent example using 100.000 doughnuts enabled the imaging of a living cell (several tens of square microns) in a few seconds [17]. This is a significant advantage over localization microscopy, which typically requires several tens of minutes to reconstruct an image.

3.3 NOVEL VARIATIONS ON THE STED PRINCIPLE

Following the introduction of STED, a number of derived techniques were developed that use doughnut or otherwise shaped beams for super-resolution imaging and tracking. The main aim of these novel variations was to overcome one of the big limitations of STED: most fluorophores can only be depleted by stimulated emission and thus require a high STED laser intensity that may result in photobleaching and eventually sample damage. Photobleaching not only occurs from the first excited singlet state but can also happen through higher excited states and triplet states as shown in Chapter 1. In this section we describe two approaches that tackle this challenge: DyMin (and its predecessor MinField) and MINFLUX. Both approaches are based on adaptive scanning strategies that reduce the overall light dose applied to a sample.

3.3.1 MINFIELD and DyMIN

The most straightforward approach to reduce the light dose on a sample and delay photobleaching is to scan only a very small (sub-diffraction limited) region such that the structures that are to be imaged never leave the center of the STED beam. This approach was dubbed MINFIELD and exploits the fact that the maxima of the doughnut beam exceed the threshold intensity by a very large factor that scales with the resolution [18]. Therefore, when recording an image by scanning the beam across the sample, each molecule in the sample is repeatedly exposed to the maxima, which exacerbates bleaching. Restricting the scanning to sub-diffraction-sized regions prevents exposure of the molecules of interest to the intensity maxima and exposes them only to the lower intensities needed for the depletion. Thereby MINFIELD largely avoids detrimental transitions to higher molecular states which not only delays bleaching by up to 100-fold but also provides access to millisecond timescales [19].

An obvious drawback of this approach is the small region on the sample that can be imaged with high resolution. DyMIN (short for Dynamic intensity MINimum) [20] is a slightly different but related strategy that minimizes exposure to high intensities except at locations where these intensities are strictly required to resolve features. This is achieved by dynamically adapting the depletion intensity (and thereby dynamically adapting the resolution) to the local structure in the sample. It thereby only uses as much depletion intensity as needed to achieve a certain preset resolution.

The concept is illustrated in Figure 3.7a, where two fluorophores spaced by less than the diffraction limit are probed. An experiment starts with a scan away from the fluorophores without doughnut-shaped depletion light (i.e., at confocal resolution). As the dye molecule is approached by the flank of the Gaussian excitation spot, the detected signal starts to increase, at which point the depletion intensity is increased just enough that the fluorophore is not detected anymore. At one of the next scan positions, the detected signal increases again, and the depletion intensity is further increased, and so on, until the desired depletion intensity (i.e., resolution) is reached. Continuing the scan and moving away from the second fluorophore on the other side, the STED power is reduced to avoid scanning the crest across the fluorophore again.

The fluorophores in DyMIN scanning therefore experience lower depletion light doses than they would in a conventional scan, since the depletion beam is not set to full power all the time. Indeed, as shown in Figure 3.7b the highest power is applied at the scan positions centered at fluorophores and their vicinity. This reduces the light dose by ~20-fold under common imaging conditions but can reach a reduction of >100-fold for sparse samples [20].

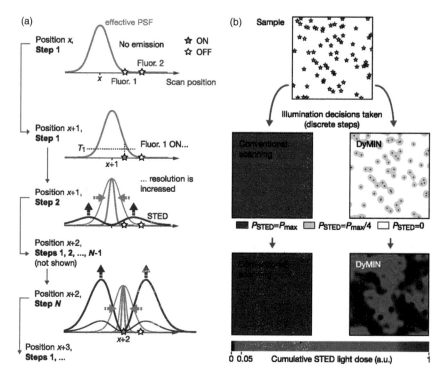

FIGURE 3.7 Nanoscopy with DyMIN adaptive illumination. (a) Concept illustrated for two fluorophores spaced less than the diffraction limit. Signal is probed at each position, starting with a diffraction-limited probing step (P_{STED} = 0, top), followed by probing at higher resolution (P_{STED} > 0). At any step, if no signal indicates the presence of a fluorophore, the scan advances to the next position without applying more STED light to probe at higher resolution. For signal above a threshold (e.g., T_1, upper middle), the resolution is increased in steps (lower middle), with decisions taken based on the presence of signal. This is continued up to a final step of P_{max} (full resolution where required). For the highest-resolution steps, directly at the fluorophore(s), the probed region itself is located at the minimum of the STED intensity profile (Bottom). (b) Simulations of conventional vs. DyMIN scanning for a sample with many fluorophores (Top). Due to different STED powers applied (Middle), different dosages result during the respective scans (Bottom). In conventional scanning, the full STED power P_{max} is applied throughout. In DyMIN, large parts of the sample are scanned with no STED light, or with STED at low powers, with the cumulative dose reduced 45-fold at the fluorophores shown. (Reproduced with permission from [20]. Copyright 2017 National Academy of Sciences.)

3.3.2 MINFLUX

In most microscopy methods, the aim is to maximize the number of measured photons on a detector, which usually leads to more rapid photobleaching for stronger signals as the photon budget for a fluorophore is limited. The concept of MINFLUX tries to accomplish the opposite: a molecule can be localized by making it coincide with the intensity zero of a doughnut-shaped excitation beam. In its simplest form this can be done by scanning a doughnut beam over a molecule and recording the fluorescence intensity as in a confocal microscope. The position of the molecule can then be identified by the minimum in signal that is obtained when the molecule is at the center of the doughnut beam. Please note that while in STED the doughnut-shaped beam is depleting the fluorescence, in MINFLUX this profile is used to excite the fluorophore.

MINFLUX can be accomplished with the setup depicted in Figure 3.8a, b. A doughnut-shaped excitation beam is scanned over the sample by a beam deflector, while recording the fluorescence intensity on a photon counter. In a typical two-dimensional MINFLUX implementation, the position of a molecule is obtained by placing the minimum of a doughnut-shaped excitation beam at a known set of spatial coordinates in the molecule's proximity. These coordinates are within a range L where the molecule is anticipated (Figure 3.8b). Probing the number of detected photons for each doughnut minimum coordinate yields the molecular position, similar to a position determination by triangulation.

The precision of the position estimate increases with the square root of the total number of detected photons and, more importantly, by decreasing the range L. For small ranges L for which the intensity minimum is approximated by a quadratic function the localization precision scales with $\sigma_{\text{MinFLUX}} \geq L/\sqrt{N}$ in the absence of noise sources other than shot noise. Comparing to the localization precision as discussed in Chapter 2 ($\sigma_{\text{loc}} \geq \sigma_0/\sqrt{N}$), the localization precision in MINFLUX can be higher if L is sufficiently small. Note that for small L the number of detected photons is in fact minimized, so the factor $1/\sqrt{N}$ that appears in σ_{MinFLUX} increases. However, a controlled reduction of L increases the localization precision more effectively than waiting for larger numbers N of detected photons. The trick is therefore to bring the donut zero virtually to spatial coincidence with the probed fluorophore, a procedure only limited by background noise. This approach has been shown to yield a localization precision of 2–5 nm in a typical experiment, not much larger than the molecular label itself [23].

Minflux Principle

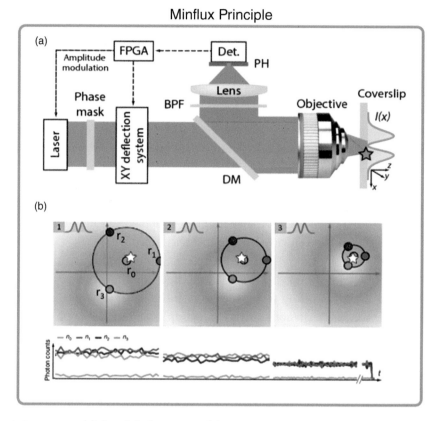

FIGURE 3.8 (a) Simplified setup used for MINFLUX. A doughnut-shaped excitation laser beam is created by a phase mask and deflected such that its central zero is sequentially placed at the four focal plane positions indicated by blue, violet, red, and yellow dots, respectively in (b). Photons emitted by the fluorescent molecule (star) are collected by the objective lens and directed toward a fluorescence bandpass filter (BPF) and a confocal pinhole (PH), by using a dichroic mirror (DM). The fluorescence photons at each position are used to extract the molecular location. Intensity modulation and deflection, as well as the photon counting, are controlled by a field-programmable gate array (FPGA). (b) The doughnut-shaped illumination pattern (green) is sequentially placed at the four positions $r_{0,1,2,3}$ indicated by blue, violet, red, and yellow dots, respectively (panel 1). The number of emitted photons $n_{0,1,2,3}$ counted for each doughnut position are used to extract the molecular location by triangulation. To improve the localization precision, the scan range of the doughnuts can be reduced iteratively until the desired localization precision is attained (panels 2–3).   (*Continued*)

Applications

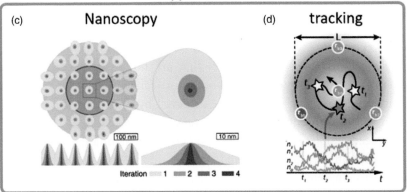

FIGURE 3.8 (CONTINUED) (c) Basic application modalities of MINFLUX. Imaging: A nanoscale object features molecules whose fluorescence switches between bright and dark states, such that only one of the molecules is on within the detection range. Iterative placement of the doughnut beam at the location of the bright molecule is achieved by minimizing the photon flux, leading to localization of the emitter. The molecule switches back to a dark state, after which the procedure is repeated for the next molecule in the bright state. (d) Nanometer-scale (short-range) tracking: The same procedure can be applied to a single emitter that moves within the localization region of size L. As the emitter moves, different fluorescence ratios are observed that allow the localization. (Panel (a) and (d) reproduced with permission under a Creative Commons CC BY-NC-ND 4.0 license from [21]. Copyright 2018 The Authors. Panel (b) and (c) reproduced with permission from [22]. Copyright 2020 Springer Nature.)

Several imaging modalities have been demonstrated using MINFLUX. Starting with nanoscopic imaging, herein a nanoscale object features molecules whose fluorescence can be depleted or switched off, such that only one of the molecules is on within the detection range. After localization of this molecule a stochastic process switches another fluorophore using mechanisms already discussed in Chapter 2. The sudden switch-on of another fluorophore is detected by an abrupt change in the ratio between the different signal intensities from the different doughnut positions (see Figure 3.8c). Repeating this process for all fluorophores near the doughnut beam results in a reconstruction, where larger areas can be imaged by moving the doughnut beam to another sample region.

In another modality, nanometer-scale (short-range) tracking employs the same procedure for a single emitter that moves within the localization region of size L. As the emitter moves, different fluorescence ratios are observed that allow the localization. As an example, MINFLUX has recently been used to probe molecular movements of a few nanometers at a temporal resolution of 400 μs [21]. This was used to probe thermal fluctuations of single DNA strands on timescales that are not easily accessible using other microscopy approaches. Beyond short-range tracking, micron-scale (long-range) tracking can again be achieved by displacing the doughnut beam as the fluorophore moves by distances larger than L.

A recent implementation of MINFLUX on a standard optical microscope used an iterative algorithm to further improve the localization precision. This implementation uses a hexagonal grid of scanning positions (rather than a triangular one as shown in Figure 3.8b) with a mutual distance of 0.1–0.5 excitation wavelengths [24]. These positions are repeatedly probed during a measurement and serve as starting points for the localization of fluorescent markers. After initial localization of the fluorophore an iterative procedure is started in which L is reduced to quickly zoom in on the emitter. While reducing L, the power in the doughnut beam is increased to keep up the fluorescence rate, resulting in the localization of an emitter with 1 nm precision in 30 ms. A more extensive discussion of MINFLUX and related approaches can be found in Chapter 7.

3.4 APPLICATION EXAMPLES IN MATERIALS SCIENCE

Until the 2010s the use of super-resolution microscopy was largely restricted to the biological sciences due to its ability to resolve sub-cellular structures with unprecedented resolution. Since the 2010s super-resolution microscopy is increasingly applied in the physical and materials sciences, which we highlight in this section by zooming in on applications of STED microscopy in fields like polymer science, nanomedicine, biosensing, and catalytic materials.

Nanoparticles are widely used in the fields of nanomedicine and biosensing because their large surface-to-volume ratio enables efficient functionalization with bio-active groups at length-scales similar to cellular structures and biomolecules. When introduced in a biological fluid such as blood, saliva, urine, or cell culture medium, the high concentration of biomolecules in the matrices often results in the formation of a so-called protein corona. Protein corona has been a topic of intense investigation [25] because it can render particles biocompatible, but it can also reduce functionality by overcoating the functional groups on the particle's surface.

STED microscopy has enabled the nanoscale investigation of protein corona on single nanoparticles [26], see Figure 3.9. Multi-color STED was implemented by sequentially scanning the sample with different excitation and STED laser beam wavelengths, resulting in 3-color nanoscopy of single silica nanoparticles. Abundant blood proteins (BSA, IgG, and transferrin) were labeled with different dyes, and their mixture was adsorbed onto the particles. Three-dimensional STED imaging then enabled the quantification of the relative content of these proteins in the corona. It was found that the composition of the corona differs strongly from particle-to-particle, and that it is influenced by the type of surface coating. Additionally, the nanoscopic resolution of STED revealed that the smallest protein (BSA) has the highest tendency to penetrate into porous particles, whereas the largest proteins were mainly found on the particle surface. Subsequent STED imaging of the interaction of nanoparticles

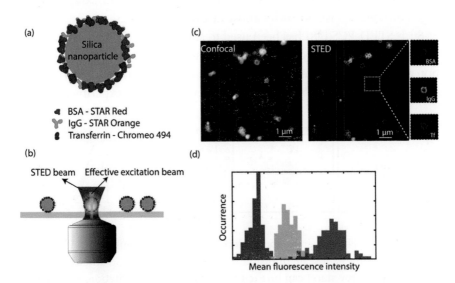

FIGURE 3.9   Workflow of STED microscopy of protein corona on nanoparticles. (a) Scheme of fluorescently labeled protein corona formation on a silica nanoparticle. (b) STED measurement of immobilized nanoparticle–protein corona complex. (c) Comparison of diffraction-limited confocal and super-resolution STED images of the same field of view and zoom-in STED images of the same particles with BSA in red, IgG in green, and Tf in blue channels, respectively. (d) Illustrative histogram (not real data) showing the mean fluorescence intensities of single particles in different color channels with matching color codes as in (c). (Reproduced with permission under a Creative Commons CC-BY license from [26]. Copyright 2022 The Authors.)

with cells or tissue could enable the correlation of protein corona distribution with cellular uptake [27].

In addition to nanoscale imaging, the focused spot of a microscope has also been used to study the diffusion of fluorescently labeled objects through the focus of a laser beam using fluorescence correlation spectroscopy (FCS). Timetraces of the fluorescence intensity then reveal a series of fluorescence bursts that can be analyzed by autocorrelation analysis. The resulting autocorrelation function (ACF) is often approximated as an exponentially decaying function and typically yields two values that can be quantified: (1) the characteristic decay time of the ACF depends on the diffusion constant of the object under study and the size of the focal spot, and (2) the plateau of the ACF at zero delay time depends on the time-averaged number of objects present in the focal spot. Both quantities depend on size of the focal spot, implying that the very small effective focal volume of a STED microscope enables the study of samples with high concentrations and with nanoscale diffusion modes.

Such STED-FCS [28] has been used to study nanoscale diffusion in synthetic membranes. By focusing the STED laser spot (consisting of the excitation and doughnut beams) onto a synthetic membrane (Figure 3.10) results in a spot size that is tunable by the STED laser power. Such experiments using

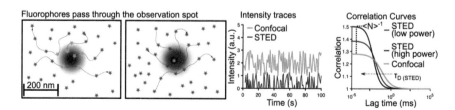

FIGURE 3.10 Fluorescence correlation spectroscopy combined with STED offers to measure nanoscale diffusion modes. Principle of FCS and STED-FCS. Fluorophores (grey stars) can be excited by the confocal excitation beam (blue, emitting molecules are depicted as green stars). When the confocal excitation beam is overlaid with a STED beam, the resulting observation spot is smaller. The recorded intensity profiles in confocal (green) and STED (magenta) illustrate lower counts for STED illumination as fewer molecules are excited than in the confocal case. Autocorrelation of the intensity traces results in the correlation curves which reveal the underlying dynamics. The transit time τ_D which is the average time a molecule needs to cross the observation spot is larger for the big confocal observation spot and shorter for the smaller STED observation spot (scaling with STED laser power). (Reproduced with permission from [31]. Copyright 2022 Springer Nature.)

spot sizes ~70-fold below the diffraction barrier revealed that unlike phosphoglycerolipids, sphingolipids and glycosylphosphatidylinositol-anchored proteins are transiently (~10–20 ms) trapped in cholesterol-mediated molecular complexes dwelling within <20-nm diameter areas [29]. Such confinement would not be quantifiable with diffraction limited beams due to the large mismatch between the focal spot size and the confinement region.

Instead of parking the beam on a sample, scanning STED-FCS [30] records FCS data across a linear or circular trajectory along which the STED beam is scanned. This enabled the mapping of membrane dynamics on a several-micrometer-long scanning trajectory by a single 10-second measurement, with a maximum spatial resolution of 60 nm and sub-millisecond temporal resolution. These measurements revealed that the diffusion speed, as well as trapping characteristics of lipids in the synthetic membrane, strongly changes in space and time across the membrane, independent of their preference for liquid-ordered or -disordered membrane environments, all in all contributing to an updated picture of plasma membrane organization.

In the field of photovoltaics, STED microscopy has enabled materials characterization on length scales that are otherwise inaccessible. Organic light-emitting diodes (OLEDs) consist of a stack of polymeric materials connected to two electrodes (Figure 3.11) that allow for electron injection

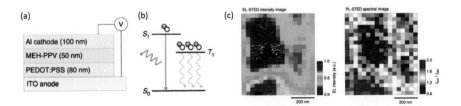

FIGURE 3.11 (a) Cross section of an OLED device consisting of a polymer bilayer (PEDOT: PSS for hole injection with MEH-PPV on top for light emission) between two electrodes (ITO on bottom, Al on top). A voltage can be applied. (b) Energy-level diagram of exciton formation in the polymer film. When the ground state S_0 is stimulated electrically into an excited singlet state S_1, spin statistics imply that uncorrelated electrons and holes pair to form three triplet (T_1) states for every singlet state. However, they relax thermally (grey arrows) leaving ~25% of singlet excitons that emit light (green arrow). (c) Intensity-based and spectral imaging of photovoltaic materials. The ratio of emission intensities at 600 nm and 580 nm is representative of the chain density (blue = low density, red = high density). (Adapted with permission under a Creative Commons CC-BY license from [32]. Copyright 2016 The Authors.)

resulting in electroluminescence [32]. STED imaging under operating conditions revealed nanoscopic defects in the polymer films on length-scales of 50 nm, inaccessible by diffraction-limited optics. By recording not only the fluorescence intensity but also its spectrum (using a regular spectrometer) provided information on the local packing density of the polymer chains whose emission spectrum red-shifts for higher packing densities. As shown in Figure 3.11c the ratio of emission intensities at 600 nm and 580 nm is strongly correlated to the intensity of electroluminescence, hinting at the influence of chain packing density on the OLED efficiency.

REFERENCES

1. S. W. Hell and J. Wichmann, "Breaking the diffraction resolution limit by stimulated emission: stimulated-emission-depletion fluorescence microscopy," *Opt. Lett.* **19**, 780 (1994).
2. T. A. Klar and S. W. Hell, "Subdiffraction resolution in far-field fluorescence microscopy," *Opt. Lett.* **24**, 954 (1999).
3. T. A. Klar, S. Jakobs, M. Dyba, A. Egner, and S. W. Hell, "Fluorescence microscopy with diffraction resolution barrier broken by stimulated emission," *Proc. Natl. Acad. Sci.* **97**, 8206–8210 (2000).
4. S. Jeong, J. Widengren, and J.-C. Lee, "Fluorescent probes for STED optical nanoscopy," *Nanomaterials* **12**, 21 (2021).
5. G. Vicidomini, P. Bianchini, and A. Diaspro, "STED super-resolved microscopy," *Nat. Methods* **15**, 173–182 (2018).
6. D. Wildanger, J. Bückers, V. Westphal, S. W. Hell, and L. Kastrup, "A STED microscope aligned by design," *Opt. Express* **17**, 16100 (2009).
7. M. Reuss, J. Engelhardt, and S. W. Hell, "Birefringent device converts a standard scanning microscope into a STED microscope that also maps molecular orientation," *Opt. Express* **18**, 1049 (2010).
8. M. P. Clausen, S. Galiani, J. B. D. L. Serna, M. Fritzsche, J. Chojnacki, K. Gehmlich, B. C. Lagerholm, and C. Eggeling, "Pathways to optical STED microscopy," *NanoBioImaging* **1**, (2014).
9. K. I. Willig, B. Harke, R. Medda, and S. W. Hell, "STED microscopy with continuous wave beams," *Nat. Methods* **4**, 915–918 (2007).
10. V. Westphal and S. W. Hell, "Nanoscale resolution in the focal plane of an optical microscope," *Phys. Rev. Lett.* **94**, 143903 (2005).
11. B. Harke, J. Keller, C. K. Ullal, V. Westphal, A. Schönle, and S. W. Hell, "Resolution scaling in STED microscopy," *Opt. Express* **16**, 4154 (2008).
12. N. T. Urban, K. I. Willig, S. W. Hell, and U. V. Nägerl, "STED nanoscopy of actin dynamics in synapses deep inside living brain slices," *Biophys. J.* **101**, 1277–1284 (2011).
13. B. R. Patton, D. Burke, R. Vrees, and M. J. Booth, "Is phase-mask alignment aberrating your STED microscope?," *Methods Appl. Fluoresc.* **3**, 024002 (2015).

14. B. Harke, C. K. Ullal, J. Keller, and S. W. Hell, "Three-dimensional nanoscopy of colloidal crystals," *Nano Lett.* **8**, 1309–1313 (2008).

15. J. Heine, C. A. Wurm, J. Keller-Findeisen, A. Schönle, B. Harke, M. Reuss, F. R. Winter, and G. Donnert, "Three dimensional live-cell STED microscopy at increased depth using a water immersion objective," *Rev. Sci. Instrum.* **89**, 053701 (2018).

16. S. J. Sahl and S. W. Hell, "High-resolution 3D light microscopy with STED and RESOLFT," in *High Resolution Imaging in Microscopy and Ophthalmology*, J. F. Bille, ed. (Springer International Publishing, 2019), pp. 3–32.

17. A. Chmyrov, J. Keller, T. Grotjohann, M. Ratz, E. d'Este, S. Jakobs, C. Eggeling, and S. W. Hell, "Nanoscopy with more than 100,000 'doughnuts,'" *Nat. Methods* **10**, 737–740 (2013).

18. F. Göttfert, T. Pleiner, J. Heine, V. Westphal, D. Görlich, S. J. Sahl, and S. W. Hell, "Strong signal increase in STED fluorescence microscopy by imaging regions of subdiffraction extent," *Proc. Natl. Acad. Sci.* **114**, 2125–2130 (2017).

19. J. Schneider, J. Zahn, M. Maglione, S. J. Sigrist, J. Marquard, J. Chojnacki, H.-G. Kräusslich, S. J. Sahl, J. Engelhardt, and S. W. Hell, "Ultrafast, temporally stochastic STED nanoscopy of millisecond dynamics," *Nat. Methods* **12**, 827–830 (2015).

20. J. Heine, M. Reuss, B. Harke, E. D'Este, S. J. Sahl, and S. W. Hell, "Adaptive-illumination STED nanoscopy," *Proc. Natl. Acad. Sci.* **114**, 9797–9802 (2017).

21. Y. Eilers, H. Ta, K. C. Gwosch, F. Balzarotti, and S. W. Hell, "MINFLUX monitors rapid molecular jumps with superior spatiotemporal resolution," *Proc. Natl. Acad. Sci.* **115**, 6117–6122 (2018).

22. K. C. Gwosch, J. K. Pape, F. Balzarotti, P. Hoess, J. Ellenberg, J. Ries, and S. W. Hell, "MINFLUX nanoscopy delivers 3D multicolor nanometer resolution in cells," *Nat. Methods* **17**, 217–224 (2020).

23. F. Balzarotti, Y. Eilers, K. C. Gwosch, A. H. Gynnå, V. Westphal, F. D. Stefani, J. Elf, and S. W. Hell, "Nanometer resolution imaging and tracking of fluorescent molecules with minimal photon fluxes," *Science* **355**, 606–612 (2017).

24. R. Schmidt, T. Weihs, C. A. Wurm, I. Jansen, J. Rehman, S. J. Sahl, and S. W. Hell, "MINFLUX nanometer-scale 3D imaging and microsecond-range tracking on a common fluorescence microscope," *Nat. Commun.* **12**, 1478 (2021).

25. P. L. Latreille, M. Le Goas, S. Salimi, J. Robert, G. De Crescenzo, D. C. Boffito, V. A. Martinez, P. Hildgen, and X. Banquy, "Scratching the surface of the protein corona: challenging measurements and controversies," *ACS Nano* **16**, 1689–1707 (2022).

26. Y. Wang, P. E. D. Soto Rodriguez, L. Woythe, S. Sánchez, J. Samitier, P. Zijlstra, and L. Albertazzi, "Multicolor super-resolution microscopy of protein corona on single nanoparticles," *ACS Appl. Mater. Interfaces* **14**, 37345–37355 (2022).

27. M. Kokkinopoulou, J. Simon, K. Landfester, V. Mailänder, and I. Lieberwirth, "Visualization of the protein corona: towards a biomolecular understanding of nanoparticle-cell-interactions," *Nanoscale* **9**, 8858–8870 (2017).

28. E. Sezgin, F. Schneider, S. Galiani, I. Urbančič, D. Waithe, B. C. Lagerholm, and C. Eggeling, "Measuring nanoscale diffusion dynamics in cellular membranes with super-resolution STED–FCS," *Nat. Protoc.* (2019).

29. C. Eggeling, C. Ringemann, R. Medda, G. Schwarzmann, K. Sandhoff, S. Polyakova, V. N. Belov, B. Hein, C. Von Middendorff, A. Schönle, and S. W. Hell, "Direct observation of the nanoscale dynamics of membrane lipids in a living cell," *Nature* **457**, 1159–1162 (2009).

30. A. Honigmann, V. Mueller, H. Ta, A. Schoenle, E. Sezgin, S. W. Hell, and C. Eggeling, "Scanning STED-FCS reveals spatiotemporal heterogeneity of lipid interaction in the plasma membrane of living cells," *Nat. Commun.* **5**, 5412 (2014).

31. F. Schneider and E. Sezgin, "Diffusion measurements at the nanoscale with STED-FCS," in *Fluorescence Spectroscopy and Microscopy in Biology*, R. Šachl and M. Amaro, eds., Springer Series on Fluorescence (Springer International Publishing, 2022), Vol. 20, pp. 323–336.

32. J. T. King and S. Granick, "Operating organic light-emitting diodes imaged by super-resolution spectroscopy," *Nat. Commun.* **7**, 11691 (2016).

CHAPTER 4

Structured Illumination Microscopy (SIM)

Lorenzo Albertazzi and Peter Zijlstra

Eindhoven University of Technology, Eindhoven, The Netherlands

4.1 INTRODUCTION

While SMLM and STED discussed in the previous chapters bypass the diffraction limit by controlling the state of the fluorescent molecules (i.e., on/off), structured illumination microscopy enhances the resolution of optical microscopy purely in an optical way. This approach does not allow for the sub-100 nm resolutions of STED and SMLM but it nevertheless has several advantages. As SIM does not depend on the photocontrol of the fluorescent labels, it works "off the shelf" on a standard specimen without the need of special dyes, buffers, or cumbersome sample preparation. Historically the sample illumination is preferably homogenous, and much effort has been spent in the last 100 years to have a constant illumination over the largest field of view possible, from the Kohler condenser reported in 1893 [1] to the modern flat-field illumination used in SMLM [2]. SIM breaks this paradigm using a non-even illumination that generates a Moiré pattern between the structured excitation beam and the structured sample. After multiple acquisitions with different patterns, a mathematical reconstruction is applied and a super-resolution image obtained. In this chapter we will discuss the principle, the implementation, and the performances of SIM and present some representative applications in biology and material science.

4.2 SIM PRINCIPLE

While attempts to improve image resolution using patterned illumination have been reported before, the first effective SIM approach can be found in the seminal paper from Mats Gustaffson in 2000 [3]. The concept of SIM can be understood in terms of the phenomenon of Moiré fringes. Moiré patterns are the result of interference between two patterns that exhibit a different shape (e.g., different pitch, different orientation, different size). This is a ubiquitous phenomenon at the macroscale, for example, artifacts of digital photos taken of objects with a regular pattern, such as a photo of a shirt with fine stripes that resulted in the interference of the stripes with the pixel-structure of the light sensor. This can be easily visualized by overlapping two identical stripe patterns rotated by a certain angle as shown in Figure 4.1. It is clear that when simply changing the angle between the two shapes different Moiré fringes appear, with different size and geometry.

The same phenomenon can happen in the nano- and micro-scale. Particularly relevant for SIM is that Moiré patterns emerge when a fluorescent sample is illuminated with a non-homogenous illumination. In this case the two interfering objects are the illumination, controlled by the optical design of the setup, with the fluorescent sample itself. Under this condition the image acquired on the camera is the resulting Moiré pattern. As it is clear from Figure 4.1, the fringes emerging are much coarser that the original pattern generating them. From a microscopy point-of-view it means that some Moiré fringes are large enough to be observed (i.e., they are larger than the resolution of the lenses, despite the fact that the original patterns were not resolvable). Therefore, the Moiré fringes contain information about the sample that is more accessible to the microscope because the fringes are "larger than the original object". If the illumination pattern is known, the shape of the sample of interest can be

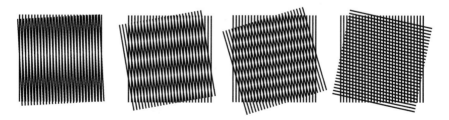

FIGURE 4.1 Moiré fringes obtained by overlapping two equal patterns consisting of parallel equidistant stripes. The different patterns are obtained by simply change the angle between the two original ones.

therefore deconvolved from the Moiré image through a mathematical procedure. This is the key to enhanced resolution of SIM: normally hidden features can be resolved by observing them under a controlled illumination pattern.

This can be more quantitatively understood by considering the object and image in Fourier space (i.e., in terms of their spatial frequencies). As discussed in Chapter 1, a microscopy image can be represented in frequency space where the low-resolution features are close to the origin in Fourier space and the high-resolution features are further away at higher frequencies (see also Figure 4.2a). Every microscopy, depending on its optical transfer function (OTF), will allow only some information to be obtained, while high resolution information is lost. Figure 4.2b shows a schematic representation of an OTF measured under standard widefield conditions and imaged under a patterned illumination (e.g., a sinusoidal pattern). Such sinusoidal modulation of the excitation pattern shifts the OTF in frequency space and thus enables retrieval of spatial frequencies that are normally lost in the microscope. Therefore, new features can be resolved, resulting in an enhanced resolution. To maximize this effect, multiple different illumination patterns are used. Figure 4.2c shows a real example of a sample imaged in frequency space with 7 different patterns (Figure 4.2c(C), showing the larger ranges of frequencies that are accessible (Figure 4.2c(D). Combining all this imagery results in an extended OTF that results in an enhanced resolution with approximatively a factor 2 compared to standard widefield resolution.

4.2.1 Practical Implementation

SIM microscopes are relatively simple in terms of optical setups. There are two main requirements: i) a known illumination pattern should be applied; ii) multiple patterns should be produced in a rapid and practical way. The procedure for a SIM image acquisition is shown in Figure 4.3a, b. The object of interest (for example. a cell) is illuminated with a striped pattern, and the same pattern is applied several times with a different angle. After recording the Moiré fringes on the camera, a mathematical reconstruction can be applied, resulting in a SIM image with enhanced resolution (Figure 4.3c, d).

SIM can be implemented in widefield, TIRF, or light-sheet illuminations by adding optical elements that generate a known pattern, typically parallel stripes with the highest possible frequency or a regular array of dots [6]. The schematic representation of a typical SIM setup is presented

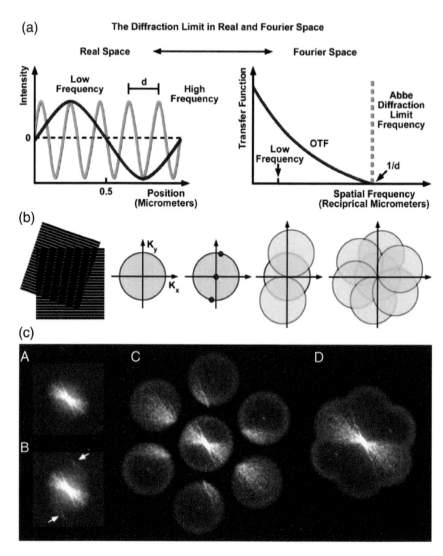

FIGURE 4.2 (a) A standard widefield image of two samples with a sinusoidally varying intensity (a low frequency in red and a high frequency in green). The optical transfer function (OTF, depicted in red in the Fourier space representation) only enables the capture of a range of frequencies limited by the diffraction limit (indicated by the green dashed line). (b) illustration of the effect of patterned illumination. The OTF is depicted as a green circle whose edges indicate the cut-off frequency shown in (a). Illuminating the sample with a sinusoidally varying pattern shifts the OTF in frequency space. Repeating this for different rotations on the pattern and computationally demixing the information results in an OTF with a higher frequency cut-off (i.e., a higher resolution image in real space). (c) example of a measured OTF for 7 different illumination patterns, together with the resulting overall OTF in D. (Adapted from [4].)

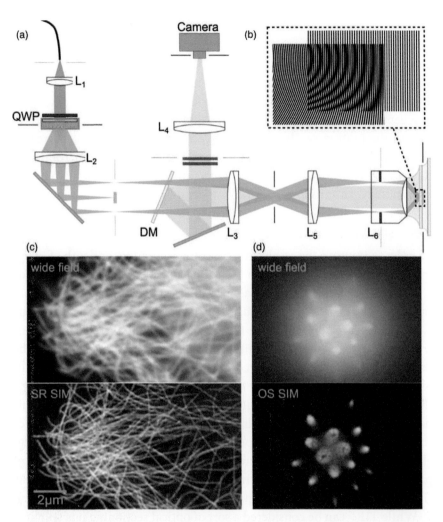

FIGURE 4.3   Acquisition procedure for a SIM image and typical comparison of SIM with classical widefield microscopy. a) schematic representation of the optical setup; b) Moiree fringes; c) 2D example of SIM resolution enhancement; d) 3D example of resolution enhancement. (Adapted from [5].)

in Figure 4.3a. The basic geometry of the system is not altered compared to a standard widefield microscope. The key difference is that a grating is placed in the excitation path, acting as an excitation mask. The grating produces three diffracted beams that are focused by a lens (L2). A blocker is used to stop the zero-order light (the central beam), resulting in two beams that are focused on the sample. The interference between these two beams creates a pattern with sinusoidal modulation that illuminates the sample. Fluorescence collection and detection is then perfectly accomplished in the same way as a widefield microscope. The grating can be easily mechanically rotated to generate patterns with different angles.

This is an inexpensive and simple implementation that guarantees robust results. However, it may lack flexibility, as the grating is fixed and can generate only a determined pattern. Moreover, the mechanical rotation of the grating can be slow and affect time resolution. To overcome these limitations, spatial light modulators (SLMs) or digital micromirrors devices (DMDs) are used in modern instruments as a faster way to generate arbitrary patterns.

4.2.2 SIM Variations

In order to improve the resolution, contrast, sectioning, or depth of imaging of SIM, several variations on the basic principles have been proposed.

TIRF-SIM has been reported as a way to improve the contrast and avoid out-of-focus signal in SIM. By using an evanescent wave excitation, the signal is measured only from the first hundreds of nm, increasing the contrast sharply. Of course, this limits the applications to thin samples or the basal membrane of cells. To overcome this, grazing incidence TIRF, TIRF-SIM has been reported (GI-TIRF) as shown in Figure 4.4a. In GI-TIRF the illumination is just inside the critical angle for SIM, generating an illumination beam parallel to the glass slide with comparable thickness of the objective depth of focus (approximately 1 μm). This is thin enough to eliminate most of the out-of-focus but deep enough to reach a significant part of the sample. This is clear from the comparison between TIRF-SIM (Figure 4.4a), widefield SIM (Figure 4.4b) and GI-TIRF SIM (Figure 4.4c).

While TIRF can be considered a sort of optical sectioning it is still limited to the first layer of the sample, and 3D imaging is not accessible. For this, 3D-SIM modalities have been implemented. Many setups have been designed but the most used approach is based on the interference between multiple beams. Figure 4.4d shows the different patterns that can be generated with one beam (top), two beams interfering (middle), and three beams interfering (bottom). Looking at the OTF functions (right) it is clear how three-beams SIM is the technique of choice for 3D imaging due to the larger extension of the OTF in the z direction and the consequent higher resolution (below 100 nm). Even more complex beam patterns can be used to further enhance the resolution, at the cost of complex instrumentation. Worth mentioning is the I5M method, where two opposing

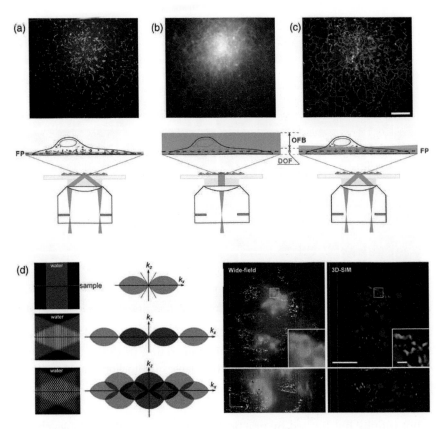

FIGURE 4.4 Optical sectioning with SIM. TIRF-SIM (a), widefield SIM (b), GI-TIRF SIM (c) and 3D-SIM (d). (Adapted from [5, 7].)

objectives are used to generate interference similarly to 4Pi microscopy, achieved 90 nm isotropic resolution in *xyz*.

A promising approach to enhance the lateral resolution of SIM is non-linear SIM. The resolution of SIM is limited by the frequency of the excitation pattern. The higher the frequency, the larger is the shift of the OTF in frequency space and therefore the higher the resolution. However, the excitation pattern is also diffraction limited and therefore limited by the objective lens and the wavelength. A possible solution to overcome this limit is to generate a non-linear response between excitation and emission. A non-linear excitation pattern will contain, together with the normal spatial frequency, harmonics with spatial frequencies that are multiples of the pattern frequency. Theoretically this allows sampling infinite spatial frequencies and therefore obtaining potentially infinite resolution. Of course, resolution

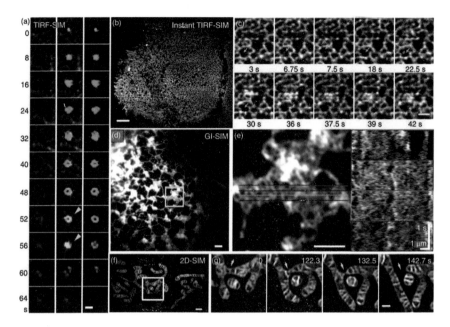

FIGURE 4.5 Biological applications of SIM. Live cell imaging of clathrin-coated endocytic vesicles (a), HRAS protein localization in time (b, c), live cell dynamics of the endoplasmic reticulum (d, e,), and time lapse of mitochondria cristae (f, g). (Adapted from [9].)

is not infinite but limited by practical factors (e.g., SNR, photobleaching) rather than by physical limits.

A simple way to achieve this is to saturate the excitation of the fluorophores (for example, applying a sinusoidal pattern with very high excitation power). Typically, this allows resolutions down to 50 nm for non-linear SIM at the cost of high photobleaching, high phototoxicity, and lower signal-to-noise [8].

4.3 APPLICATIONS

Due to the simplicity of the optical implementation and the low light dose used for illumination, SIM became rapidly popular in biology. Moreover, SIM does not need special dyes, allowing biologists to immediately apply it to existing samples (e.g., cells genetically modified with fluorescent proteins) without the need to modify existing labeling protocols or generate new genetically modified cells or organisms. While SIM features are not particularly advantageous for fixed samples that, of course, are not prone to photodamage or phototoxicity, the rapid speed and low illumination dose of SIM made it the technique of choice for fast live-cell imaging beyond the diffraction limit. Figure 4.5 shows an overview of different organelles visualized in live cells with different SIM implementations. In all cases, sub-diffraction time lapses of the structure of interest are obtained without significant cell toxicity. Of course, if a resolution below 100 nm is needed other methods are needed (e.g., STED), making live cell imaging more complicated.

Despite the fact that SIM is particularly attractive for biological applications, it also found several uses in the field of materials. The low illumination dose for example can be beneficial for photosensitive materials that cannot be irradiated with the high powers necessary for STED or SMLM. Moreover, techniques like SMLM strongly rely on tight control of the dye's photophysics and environment that is often not compatible with many materials. A striking example of the power of SIM to unveil material structure is the recent work of Ian Manners and coworkers [10].

In this work polymeric "nanoplatelets" with controlled solid and hollow 2D structure are presented. By means of seeded crystallization, the authors can control shape and size of the materials and, by sequential addition of different polymer blends, concentric multilayer materials can be obtained. This poses two characterization questions: i) how to check the size and morphology of the obtained material and ii) how to prove the differential distribution of the different polymeric building blocks. To be able to answer these questions, both nanometric resolution and multicolor in situ imaging are necessary.

It is proven that SIM is promising for studying both live cells and materials, so an obvious follow-up application is the study of material interactions in living cells. Following the position and the properties of materials during cell internalization is of great interest for biomaterials but often hampered by the limited resolution of optical microscopy or by the need to fix/freeze cells for electron microscopy and the consequent impossibility of running a time-lapse study. A particularly interesting example of SIM for the study of material-cell interactions has been recently report by Caruso and co-workers [11].

In this report the authors studied the effect of cell interactions on the structural integrity of 500 nm polymeric nanocapsules. The hypothesis is that cells can exert forces on these soft materials, resulting in capsule deformation or rupture in time. To verify this, a technique is needed that is able to perform live-cell imaging with high speed and material imaging with nanometric resolution. Thanks to 2-color SIM, the authors proved that

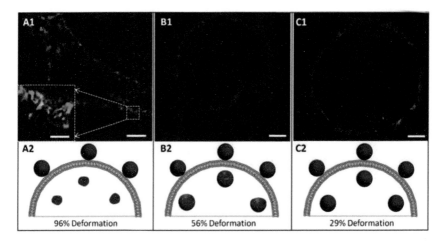

FIGURE 4.6 Polymeric capsules deformation in HeLa cells (A1-A2), Raw Macrophages (B1-B2) and THP-1 cells (C1-C2). (Adapted from [11].)

different cell types exert different forces and consequently result in different deformations: Hela cells deformed the totality of internalized capsules, while the material is almost unaffected in THP-1 cells (Figure 4.6).

4.4 CONCLUSIONS

SIM is one of the first super-resolution methods implemented since the early 2000s. While due to its limited resolution (120 nm typically) SIM may seem inferior to other methods at a first glance, its other great advantages make it the best choice in many situations. In particular, SIM is the technique of choice when fast imaging is needed, especially when a time lapse is needed due to its minimal light dose and the consequent minimal photobleaching, photodamage, and phototoxicity. Taking into account the simple instrumentation, we envision a wide use in the future to replace widefield imaging in materials and material-cell interactions.

REFERENCES

1. A. Kohler, "Ein neues Beleuchtungsverfahren für mikrophotographische Zwecke.," *Zeitschrift für wissenschaftliche Mikroskopie und für Mikroskopische Technik* **10**, 440–443 (1893).
2. K. M. Douglass, C. Sieben, A. Archetti, A. Lambert, and S. Manley, "Super-resolution imaging of multiple cells by optimised flat-field epi-illumination," *Nat. Photonics* **10**, 705–708 (2016).
3. M. G. L. Gustafsson, "Surpassing the lateral resolution limit by a factor of two using structured illumination microscopy," *J. Microsc.* **198**, 82–87 (2000).

4. J. R. Allen, S. T. Ross, and M. W. Davidson, "Structured illumination microscopy for superresolution," *ChemPhysChem* **15**, 566–576 (2014).
5. Y. Ma, K. Wen, M. Liu, J. Zheng, K. Chu, Z. J. Smith, L. Liu, and P. Gao, "Recent advances in structured illumination microscopy," *J. Phys. Photonics* **3**, 024009 (2021).
6. A. Jost and R. Heintzmann, "Superresolution multidimensional imaging with structured illumination microscopy," *Annu. Rev. Mater. Res.* **43**, 261–282 (2013).
7. Y. Guo, D. Li, S. Zhang, Y. Yang, J.-J. Liu, X. Wang, C. Liu, D. E. Milkie, R. P. Moore, U. S. Tulu, D. P. Kiehart, J. Hu, J. Lippincott-Schwartz, E. Betzig, and D. Li, "Visualizing intracellular organelle and cytoskeletal interactions at nanoscale resolution on millisecond timescales," *Cell* **175**, 1430–1442.e17 (2018).
8. M. G. L. Gustafsson, "Nonlinear structured-illumination microscopy: wide-field fluorescence imaging with theoretically unlimited resolution," *Proc. Natl. Acad. Sci. U.S.A.* **102**, 13081–13086 (2005).
9. Y. Wu and H. Shroff, "Faster, sharper, and deeper: structured illumination microscopy for biological imaging," *Nat. Methods* **15**, 1011–1019 (2018).
10. H. Qiu, Y. Gao, C. E. Boott, O. E. C. Gould, R. L. Harniman, M. J. Miles, S. E. D. Webb, M. A. Winnik, and I. Manners, "Uniform patchy and hollow rectangular platelet micelles from crystallizable polymer blends," *Science* **352**, 697–701 (2016).
11. X. Chen, J. Cui, H. Sun, M. Müllner, Y. Yan, K. F. Noi, Y. Ping, and F. Caruso, "Analysing intracellular deformation of polymer capsules using structured illumination microscopy," *Nanoscale* **8**, 11924–11931 (2016).

CHAPTER 5

Other Super-Resolution Approaches

Lorenzo Albertazzi and Peter Zijlstra

Eindhoven University of Technology, Eindhoven, The Netherlands

5.1 INTRODUCTION

Although SMLM, STED, and SIM are the most used approaches toward super-resolution imaging, other methods have also been developed. Albeit less broadly applied, each approach has its respective advantages (and disadvantages) and may be the method of choice for a specific application or in the case of instrument availability or budget restrictions. In this chapter we describe three approaches that have been reported in the past decade. We start with fluctuation-based imaging, a method that combines the switching on- and off of fluorophores and advanced image analysis approaches to improve the resolution of even very densely labeled samples in as little as 100 camera frames. We continue our discussion with image scanning microscopy, an extension of confocal microscopy that reaches a $\sqrt{2}$ increase in image resolution. We finish with near-field imaging techniques that exploit the strong confinement of light around the sharp tip of, for example, a cantilever in an atomic force microscope. By scanning the tip across a sample, it is used as a local nanometer-scale excitation source, routinely achieving resolutions of several tens of nanometers, even at near-infrared (micrometer) wavelengths.

5.2 FLUCTUATION-BASED IMAGING

Fluctuation-based imaging approaches rely on the statistical analysis of fluorescence images over time. As the name suggests, it requires a sample that is labeled with emitters that fluctuate in intensity. These emitters

DOI: 10.1201/9781003220688-5

could be dyes that blink due to triplet-state dynamics or quantum dots that blink due to the presence of trap states on their surface. Noteworthy is that these fluctuations do not have to occur between a bright and completely dark state like in SMLM but could also be caused by slight (but detectable) fluctuations in intensity or spectrum of the emission. The main difference between localization microscopy (discussed in Chapter 2) and fluctuation-based imaging is that the former is based on single-molecule detection and uses algorithms to localize individual emitters that switch on and off. This results in an approach that is more precise but at the expense of complex sample preparation, stringent requirement on the blinking behavior of the probes, or prolonged imaging times [1].

One of the first approaches to fluctuation-based imaging was reported in 2009 and dubbed Stochastic Optical Fluctuation Imaging (SOFI) [2]. SOFI uses a series of fluorescence images of a sample that is labeled with fluorophores that each have a similar brightness (see Figure 5.1). Each fluorophore generates a point spread function on the camera that does not need to be separated in space by more than the diffraction limit, as was the case for localization microscopy. At each pixel the fluorescence signal fluctuates in a time-dependent manner due to, for example, fluorescence blinking or other sources of intermittency.

Let us first find an expression for the detected fluorescence signal at position r and time t. This signal is given by $F(r,t) = \sum_{k=1}^{N} U(r - r_k)\epsilon_k s_k(t)$, where $U(r - r_k)$ is the PSF of a fluorophore at position r_k, ϵ_k is the (constant) molecular brightness and $s_k(t)$ is a function that describes the fluctuations in the brightness. The summation runs over all N fluorophores that contribute to the signal. The signal fluctuations can be analyzed by the temporal autocorrelation function of each pixel, where we assume that there are no other sources of intensity fluctuation (caused by drift, for example).

The second order autocorrelation function of the pixel intensity can be computed by multiplying the time-dependent pixel intensity fluctuations δF with a time-delayed version of itself, in other words:

$$G_2(\tau) = \langle \delta F(r, t+\tau)(\delta F(r,t)) \rangle_t$$
$$= \sum_{j,k} U(r - r_j) U(r - r_k)\epsilon_j \epsilon_k \langle \delta s_j(t+\tau)\delta s_k(t) \rangle,$$

where j and k indicate all sources of intensity (e.g., all fluorophores) that contribute to the pixel's intensity with their respective brightness and intensity fluctuation. If we assume that all intensity fluctuations are stochastic

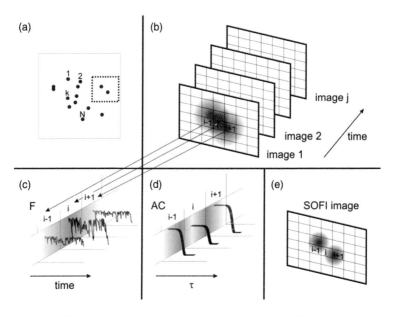

FIGURE 5.1 Principle behind super-resolution optical fluctuation imaging (SOFI). (a) Emitter distribution in the object plane. Each emitter exhibits fluorescence intermittency, which is uncorrelated with the others. (b) Magnified detail of the dotted box in (a). The signal from the emitter fluorescence distribution is convolved with the systems PSF and recorded on a camera. Two neighboring emitters, for example, cannot be resolved because of the optical diffraction limit. The fluctuations are recorded in a movie. (c) Each pixel contains a time trace, which is composed of the sum of individual emitter signals, whose PSFs fall (partly) onto the pixel. (d) The second-order correlation function is calculated from the fluctuations for each pixel. (e) The SOFI intensity value assigned for each pixel is given by the integral over the second-order correlation function. The second-order correlation function is proportional to the squared PSF, thus increasing the resolution of the imaging system by a factor of $\sqrt{2}$. (Adapted with permission from [2]. Copyright 2009 National Academy of Sciences.)

(i.e., the emission of different fluorophores is not correlated and all terms with $j \neq k$ vanish) we find

$$G_2\left(r, \tau\right) = U^2\left(r - r_k\right)\epsilon_k^2 \left\langle \delta s_k\left(t + \tau\right)\delta s_k\left(t\right)\right\rangle.$$

The second order autocorrelation function thus scales with the square of the PSF, weighted by each emitter's squared brightness and molecular correlation function. The value of $G_2(r, \tau)$ at a certain lag-time τ then defines a SOFI image, the only difference between the images being the weighting of the squared PSF with the molecular correlation function [2]. The intensities of a SOFI image therefore do not represent the fluorescence signal but rather its brightness and degree of correlation. In practice the integral over the second-order correlation function is plotted, essentially replacing the original PSF with the square of the PSF. If we assume that the PSF is a Gaussian function, this results in an increased resolution by a factor of $\sqrt{2}$ because the standard deviation of a squared Gaussian is reduced by $\sqrt{2}$ compared to the original Gaussian.

For further resolution enhancement, it is possible to compute SOFI of higher orders ($n \geq 3$). This approach does not auto-correlate the signal; instead, it computes further statistical descriptors of the fluorescence dynamics called temporal cumulants, which are similar to the statistical moments [3]. Higher-order auto-cumulant SOFI improves the resolution by a factor of $\sqrt{n}$, where n is the order of SOFI (Figure 5.2). In theory, a fourth-order SOFI generates a two-fold improvement in resolution, and a 16th-order SOFI can achieve a four-fold resolution increase.

A limitation of SOFI is that the image pixel size is constant between the input diffraction-limited dataset and the super-resolved image. This issue imposes a boundary to the achievable resolution, where sampling at the Niquist-Shannon rate (two pixels per diffraction-limited spot) provides a resolution improvement of $\sqrt{2}$. For very bright samples such as quantum dots, the pixel size of the diffraction-limited dataset can be oversampled (>2 pixels covering the full width of the PSF), and higher orders of SOFI will deliver images with enhanced spatial resolution (See Figure 5.2g). In addition, more advanced algorithms exploiting spatio-temporal cross-cumulants have been demonstrated in which the resolution scales linearly with the cumulant order (i.e., with n rather than $\sqrt{n}$ as described above), albeit at the expense of mathematical complexity [4].

SOFI does not demand the use of controllably blinking fluorophores due to the analysis of fluctuations over time. Nevertheless, better blinking kinetics from optimized dyes will result in cleaner autocorrelation functions and therefore a more significant improvement in resolution in the super-resolved image [5]. The number of frames required to reconstruct a super-resolved image using SOFI depends on the SOFI variant but is typically 3000 to 5000 frames for the third order and >10,000 frames for the fourth order [6].

A related approach is called super-resolution radial fluctuation (SRRF) that reconstructs a super-resolution image with only 100 frames and achieves a resolution of ~60 nm on a range of microscopes (TIRF, confocal,

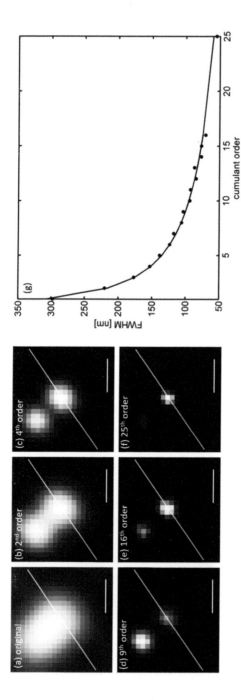

FIGURE 5.2 Higher-order SOFI images. (a–f) Selected SOFI images acquired from a movie taken with QDs deposited on a coverslip are shown. From upper left to lower right: Original image (mean intensity of all movie frames) and 2nd, 4th, 9th, 16th, and 25th orders of SOFI. The two different QDs are resolved at higher-order SOFI images. Note that the relative intensities of the two QDs vary depending on the specific blinking characteristics of each QD, which is addressed at different cumulant orders. (b) Resolution enhancement of SOFI, the full width at half maximum (FWHM) is obtained from a Gaussian fit across the dotted line profiles in (a–f). (Reproduced with permission from [2]. Copyright 2009 National Academy of Sciences.)

and epi-fluorescence). SRRF was first developed in 2016 [7] and consists of two main steps: a spatial analysis in which an algorithm constructs a so-called radiality map for each camera frame, followed by a temporal analysis of each radiality map to generate a super-resolved image (see Figure 5.3).

SRRF assumes the image is formed of point sources convolved with a point spread function (PSF) that displays a higher degree of local symmetry than the background [7]. As shown in Figure 5.3a the radiality map is generated from an interpolated camera image, resulting in an image sequence in which the pixels are sub-divided into several tens of sub-pixels. For each frame in an image sequence SRRF then calculates the local gradient along the x- and y-axis, i.e., $G_x(x, y) = \dfrac{\delta I(x, y)}{\delta x}$ and $G_y(x, y) = \dfrac{\delta I(x, y)}{\delta y}$, where $I(x, y)$ is the intensity on the camera. The gradient map can then be

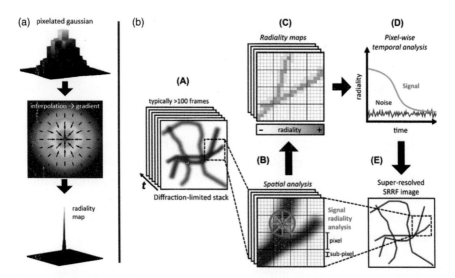

FIGURE 5.3 Principle of SRRF microscopy (a) Spatial analysis example for a pixelated Gaussian. Top: 3D surface plot of a pixelated simulated widefield PSF. Middle: surface plot of the gradient magnitude (arrows indicate direction). Bottom: 3D surface plot of the measured radiality PSF. (b) Spatial and temporal analysis of a sequence of camera frames. (A) Each image of the input sequence is (B) subpixel interpolated and on each of them a radiality measure is performed. (C) A radiality map is generated on every image from the sequence and (D) they are correlated along time in order to (E) form the reconstructed image. ([a] Reproduced with permission under a Creative Commons CC-BY 4.0 license from [7]. Copyright 2016 The Authors; [b] Reproduced with permission under a Creative Commons CC-BY 4.0 license from [6]. Copyright 2022 The Authors.)

interpolated and corrected for mechanical drift during the measurement. The gradient maps are then converted to radiality maps by calculating the degree to which the gradients converge to a point [7]. In the case of a single fluorophore this results in a continuous conical distribution with a significantly reduced full-width-half-maximum (FWHM) that can be displayed on an up-sampled pixel grid (Figure 5.3a). This radiality distribution is capable of distinguishing two Gaussian PSFs separated by ∼0.7 times the Gaussian FWHM. Because noise pixels can also contribute to the radiality maps, these are removed by weighting the radiality map by the fluorescence intensity.

This process is repeated for all frames, resulting in a stack of radiality maps that can be further denoised and enhanced in terms of contrast by temporal analysis (Figure 5.3b(A–C)).

Temporal analysis is then performed, for example, by calculating the autocorrelation function that results in a strong correlation for radiality caused by fluorescence signal, whereas noise is removed because it is not correlated between images. Note that the purpose of temporal analysis in SRRF is therefore to remove noise, whereas in SOFI such temporal analysis is used to identify fluorophore blinking. As in the case of SOFI, higher-order temporal cumulants can be calculated on the radiality map, further enhancing contrast and, in the case of low-density data sets, increasing resolution by reducing the FWHM of the radiality distribution [7]. However, compared with SOFI, the strong dependence on fluorophore brightness is largely removed because the radiality distribution is independent of signal intensity.

Figure 5.4 shows a comparison in performance between SRRF and localization microscopy as discussed in Chapter 2 for varying fluorophore densities and illumination powers. The fluorophore density was here "varied" by varying the illumination intensity since lower intensities result in fewer transitions to a dark state for Alexa Fluor 647. By selecting three different laser illumination intensities, data sets with varying fluorophore density can be produced. In the case of many overlapping fluorophores, the local radial symmetry is reduced and consequently the magnitude of the radiality is diminished; however, a conical distribution with a reduced FWHM is still observed parallel to elongated structures, which enables their imaging with super-resolution. At such ultrahigh fluorophore densities, the multi-emitter fitting procedures failed but SRRF can still extract the structure and achieve a resolution of ∼130 nm for the lowest illumination intensity shown in Figure 5.4. For the high illumination power, the datasets mainly

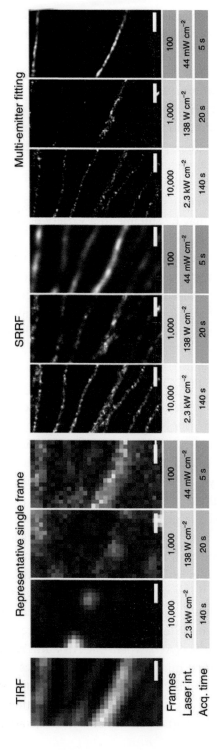

FIGURE 5.4   SRRF analysis for different fluorophore densities. Fixed microtubules were labeled with Alexa Fluor 647 and imaged with different laser intensities to produce different length data sets of varying fluorophore densities. Number of frames in data set, on-sample laser intensity, and total acquisition time shown underneath images. The same region of the sample was imaged under each set of conditions. Far left: TIRF image of region. Left: representative single frames from acquired data sets. Middle: reconstructions from SRRF. Right: reconstructions from multi-emitter fitting with maximum likelihood estimation as described in Chapter 6. Scale bars, 500 nm. (Reproduced with permission under a Creative Commons CC-BY 4.0 license from [7]. Copyright 2016 The Authors.)

contained isolated blinking fluorophores (much like a STORM, PALM, or PAINT dataset) in which SRRF achieved a resolution of ~30 nm, similar to localization microscopies.

Notably, the performance of SRRF on high-density data sets allows the production of a super-resolution image with as few as 100 raw frames (achievable in <1 s using current cameras) whereas emitter fitting would require thousands of frames or more. The algorithms to perform SRRF have been implemented on graphics processing units [8] and even in commercially available cameras [9] yielding automated and real-time super-resolution microscopy.

Overall, SOFI and SRRF provide a simpler alternative to single-molecule localization microscopy and may be suitable in a variety of applications where speed or live cell imaging are necessary. Notably, they do not need special instrumentation and can therefore be accessible to a vastly larger audience. Of course, this comes at the price of a decreased resolution, as well as the loss of the single molecule information contained in SMLM data (e.g., molecular counting). Moreover, as with any image-analysis-based method, it is important to carefully assess the input and output data to avoid artifacts.

5.3 IMAGE SCANNING MICROSCOPY

As discussed in Chapter 1 a conventional confocal microscope is based on two ingredients: a focused laser beam that is raster scanned over the sample, and a confocal pinhole placed before the single-pixel detector that rejects out-of-focus light. With this, true three-dimensional images can be recorded because only light emanating from a plane near the laser focus passes through the pinhole. In a well-adjusted confocal microscope, the confocal pinhole is chosen large enough so nearly all the light from the sample can pass. This assures maximum signal reaching the detector, resulting in an optimum signal-to-noise ratio. In that case, the lateral resolution of the microscope is given by Abbe's famous formula FWHM = λ/2NA, where we ignore the difference in wavelength between excitation and emission light. The scalings of the lateral and axial resolution are shown in Figure 5.5 as a function of the pinhole size. Often the pinhole size is depicted in so-called Airy units (AU), where one Airy unit represents a pinhole size equivalent to the first minimum of the Airy disk. At a pinhole size of 1 AU the lateral resolution is close to the diffraction limit, whereas the axial resolution is substantially improved because the pinhole blocks out-of-focus light (optical sectioning). It can also be seen that, by decreasing

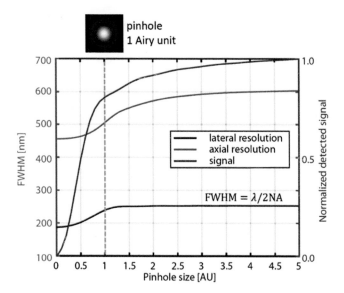

FIGURE 5.5   Influence of the pinhole diameter on the lateral (black) and the axial (blue) resolution as well as the detected signal (red). Calculations are performed for an NA 1.4 oil-immersion objective lens (λ_{exc} = 640 nm, λ_{det} = 680 nm, n = 1.518). Increasing the pinhole size increases the detected signal, but also lowers the achievable resolution. Often pinholes with a size of 1 AU are utilized in a confocal microscope, as at this size sufficient signal is collected while the optical sectioning capability is maintained and the lateral resolution is close to Abbe's diffraction limit. (Adapted with permission under a Creative Commons CC-BY 4.0 license from [10]. Copyright 2020 The Authors.)

the pinhole size further, the lateral resolution can be improved further but at the expense of a large fraction of the signal. In other words, higher-resolution images result in poor signal-to-noise ratios, a phenomenon that can be overcome by so-called image scanning microscopy.

In contrast to a conventional confocal-laser scanning microscope, which employs a single-pixel or point detector that detects all fluorescence emanating from the pinhole, Shepard already proposed in 1988 to image the fluorescence light on a CCD camera [11]. Thus, for each position of the excitation spot, a whole image is recorded, so that the method was called image scanning microscopy or ISM. Thus, after one scan with the laser focus, one did not record just a single scan image, but as many scan images as the array detector has pixels. Due to the small pixel size, the spatial resolution of these scan images corresponds to that of a confocal microscope with a zero-size pinhole. However, there are no light losses because all collected fluorescence light will hit somewhere on the array detector.

The first experimental implementation of the approach was in 2010 by Enderlein et al. [12], who used a camera as the detector. A more recent commercial implementation is shown in Figure 5.6 [13], where the beam scanning mirror is not depicted in the schematic for clarity. The displayed implementation contains two detection paths to directly compare regular confocal imaging (top beam path, with a single-pixel detector that integrates all fluorescence photons) and image scanning microscopy (where the pinhole has been removed and the single-pixel detector replaced by an array of photomultiplier tubes). As shown in Figure 5.6b, the size of each

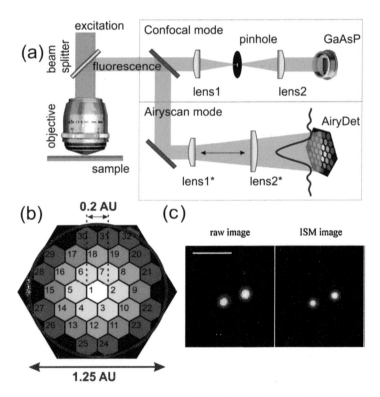

FIGURE 5.6 Airyscan microscope principle and configuration: (a) scheme of confocal and Airyscan detection paths; (b) 32 detection element arrangement in the Airyscan detector design; (c) representative images of a 100-nm fluorescent bead imaged using confocal imaging with an aperture of 1 AU, and the ISM image. Scale bar is 1 μm. (Reproduced with permission under a Creative Commons CC-BY 4.0 license from [14]. Copyright 2017 The Authors.)

pixel is equivalent to 0.2 AU, thereby improving the spatial resolution by a factor ~1.4 over a traditional confocal microscope with a pinhole of 1 AU (as shown in Figure 5.5), while there is no longer sacrifice in signal strength because all fluorescence photons are collected by the array detector.

The main task is then to combine all scan images into a single high-resolution image. Simply adding all images into a final summed image will obviously not work because each pixel has a slightly different viewing

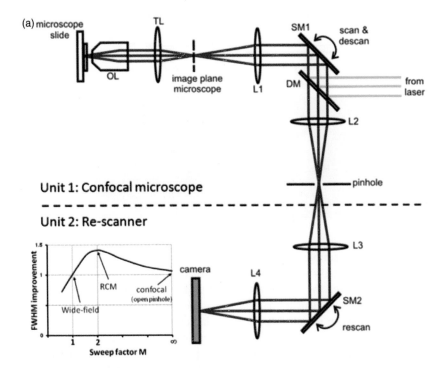

FIGURE 5.7 The Re-scan Confocal Microscope (RCM) (a) consists of two units: 1) a standard confocal microscope with a set of scanning mirrors that have a double function: scanning the excitation light and de-scanning the emission light, and 2) a re-scanning unit that "writes" the light that passes the pinhole onto the CCD-camera. The ratio of angular amplitude of the two scanners, expressed by the sweep factor M, changes the properties of the re-scan microscope (see inset in a). For $M = 1$ the microscope has the lateral resolution of a wide-field microscope, defined by the diffraction limit. For $M = 2$ the RCM performs best concerning resolution. Even with a wide-open pinhole the resolution is $\sqrt{2}$ times improved, which makes the system much more photo-efficient compared to conventional confocal microscopes with similar resolution (that should have pinhole < 1 Airy unit). For large values of M the system converts to a confocal microscope with open pinhole. *(Continued)*

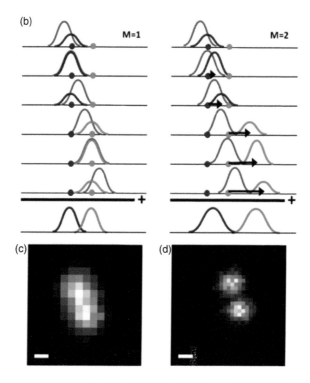

FIGURE 5.7 (CONTINUED) The Re-scan Confocal Microscope (RCM) (b) The concept of RCM is simple: For $M = 1$ the two scanners compensate each other and two point-objects (red and green dots) are projected on the camera without extra magnification. When the sweep-factor is set to $M = 2$, the extra sweep (indicated with black arrows) will smear out the spots on the camera by a factor of $\sqrt{2}$. Since the distance between the objects is 2 times larger, the relative width of the spots is reduced by a factor of $\sqrt{2}$. This resolution improvement is clearly visible by comparing (c) and (d). Two 100 nm beads are positioned 250 nm from each other and cannot be resolved by $M = 1$ (c) but can easily be separated by RCM with $M = 2$ (d). FWHM in these two configurations was 255 nm and 185 nm, respectively. Scale bars are 100 nm. (Reproduced with permission under a Creative Commons CC-BY 4.0 license from [16]. Copyright 2013 The Authors.)

angle (parallax) onto the sample [15]. Essentially, since in ISM every detector pixel sees the object at a slightly different parallax, rather than summing the detected signal from all pixels as in a standard confocal microscope, the contributions from different pixels are first shifted to account for the parallax and then summed. This process is termed 'pixel-reassignment'. Experiments and simulations have shown that the optimum process of pixel reassignment virtually shifts each pixel by half its distance from the optical axis. Subsequently, the obtained rescaled images are accumulated by adding them to a final image at the position corresponding to the focus position in the sample [15]. The results of such pixel reassignment is shown in Figure 5.6c where the resolution of the image has improved by ~1.4x, similar to the improvement expected for a standard confocal microscope with an infinitely small pinhole.

The above approach uses a computer (post-processing) for pixel-reassignment. A different implementation uses an additional scanner for hardware-based pixel reassignment. One possible all-optical implementation is schematically shown in Figure 5.7, which is termed rescan confocal microscopy [16]. Instead of shrinking the image at each focus position and then placing it back in the final image at the corresponding focus position (as done in Figure 5.6), one can alternatively maintain the image size and place them with double distance to each other on the camera. The setup shown in Figure 5.7a displays a regular confocal microscope (top half of the schematic) that is modified with a so-called re-scanner that scans the beam at twice the original amplitude. The result is that the detected emission spots remain the same size but are spread by twice the distance. By rescaling the final image one then obtains an improvement in the resolution of ~1.4x in an all-optical way. Figure 5.7b-d show the dependence of the resolution improvement on the amplitude of the re-scanner (the sweep factor). For a sweep-factor of 1 the microscope acts as a regular widefield microscope, while for very large sweep-factors the microscope acts as a confocal microscope with open pinhole. The advantage of the re-scan confocal approach is that it does not require software processing for pixel reassignment and that it can be incorporated as an extension on existing confocal microscopes.

Thanks to the commercially available solutions and its simplicity ISM is growing in popularity among microscopists. ISM provides a limited but significant enhancement in resolution compared to standard confocal microscopy without any other loss. It is therefore ideal when a small enhancement of resolution is beneficial while the sample preparation, labeling, and analysis cannot be altered.

5.4 NEAR-FIELD IMAGING

Visualizing single atoms with a typical size <0.5 nm on a solid surface has been a longstanding dream for researchers [17] and was finally fulfilled by the invention of scanning tunneling microscopy (STM) in 1981 by Rohrer, Binnig, and Gerber at IBM Zürich [18]. STM uses a sharp metal tip to collect tiny currents from the sample surface, based on the quantum

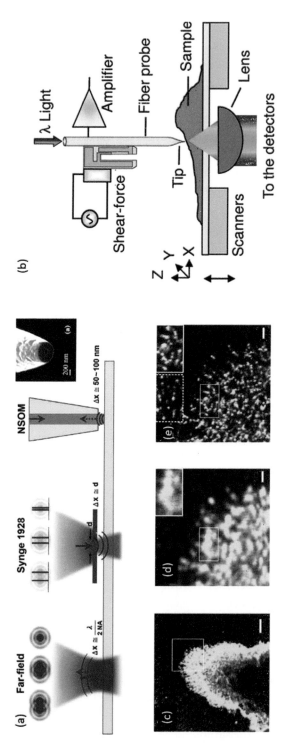

FIGURE 5.8 Near-field scanning optical microscopy (NSOM). (a) The resolution of far-field optics is dictated by the Abbe's diffraction limit. Synge proposed in 1928 reducing the size of the illumination spot using an aperture, in which case the resolution is dictated by the size of the aperture and the proximity to the sample. Current implementations use a fiber with an aperture on the order of tens of nanometers for illumination. (b) The aperture is used to guide the light to the sample, where the aperture is kept at a nanometric distance from the sample using force feedback. A far-field objective lens is often used to collect the signal, but it can also be collected through the same fiber. (c–d) Confocal images of the fluorescently labeled membrane protein GM1. (e) The same area imaged using NSOM. Scale bars 5 μm in (c) and 1 μm in (d–e). ([a] Adapted with permission under a Creative Commons CC-BY 4.0 license from [23]. Copyright 2020 The Authors; [b] Reprinted with permission from [24]. Copyright 2012 American Chemical Society; [c]–[e] Reprinted with permission from [25]. Copyright 2010 National Academy of Sciences.)

mechanical effect termed tunneling. The resolution of this method is largely determined by the size of the tip, which could be made atomically sharp to resolve single atoms on the surface.

Soon after the invention of STM, Binnig et al. were able to map the tip–sample interaction by atomic forces instead of the tunneling current, which led to the birth of the atomic force microscopy (AFM) [19]. AFM is able to image almost any type of surface, unlike STM that requires conducting or semiconducting samples. In both AFM and STM the basic principle of operation is similar to a record player, where a tip senses the surface topography of the sample via tip-scanning.

The invention of the AFM was soon followed by the development of near-field optical microscopy, where a tip or a fiber is kept in close proximity to the sample surface in order to probe its optical properties. Recall that for far-field optical imaging the ability to resolve two features is determined by Abbe's diffraction limit providing an optical resolution of $\Delta x = \lambda/2\mathrm{NA}$. The original theoretical concept of near-field imaging was proposed by E. H. Synge in 1928 [20]. He proposed to mill a small aperture (diameter of <10 nm) in a metal film and use the transmitted light to illuminate a sample. However, his idea could not be practically implemented for a long time due to difficulties in nanofabrication and positioning the aperture close to the film. Developments that led to the introduction of AFM however solved these challenges, paving the way to the development of near-field microscopy, which was first reported in 1984 [21, 22].

Near-field scanning optical microscopy (NSOM) is the first approach for measuring near-field optical responses using an optical fiber probe (Figure 5.8a) [23]. A metal-coated optical fiber probe has a nano-aperture with a diameter of ~50–100 nm that is used to image a sample by scanning the fiber with respect to the sample. The fiber probe is kept in close proximity to the sample using the force-feedback loops commonly used in AFM. One possible implementation is shown in Figure 5.8b, where the fiber probe is connected to a tuning fork whose resonance is tracked in an electronic circuit. Shear forces exerted on the fiber tip by the underlying sample modulate the resonance of the tuning fork. Monitoring and stabilizing the resonance via a feedback loop and an amplifier, thereby enabling the fiber probe to track the sample surface at a constant distance (i.e., a constant force). In the implementation in Figure 5.8b the image is formed by exciting the sample through the narrow fiber opening and detecting the light emitted by the sample, using a regular lens. Other implementations excite the sample through a lens and collect emission through the fiber, or guide both excitation and emission through the fiber.

With a resolution far beyond the diffraction limit, many groundbreaking results were reported in fields ranging from physics to chemistry and biology [23]. One example is shown in Figure 5.8c–e which displays images of the fluorescently labeled cell-membrane molecule GM1. The image collected using NSOM has a distinctly higher resolution because the sample is excited through the narrow fiber opening. Nevertheless, the application range of NSOM had been limited because the transmission intensity through the fiber probe is inversely proportional to the fourth power of the aperture size. This means that increasing resolution by reducing the aperture size results in severe loss of excitation or detection efficiency due to the high losses for apertures smaller than ~100 nm.

To overcome these constraints of an optical fiber probe, an inverse approach called apertureless NSOM (a-NSOM) or scattering-based near-field optical microscopy (s-SNOM), was developed (Figure 5.9). In s-SNOM, light of the visible, infrared, or terahertz spectral range is focused onto the tip of a standard atomic force microscope (AFM) probe using a parabolic focus mirror. The tip herein further concentrates the excitation light into highly confined and enhanced profiles at the very tip apex due to the "lightning-rod effect" (Figure 5.8a and [26]). Depending on the material properties of the tip, the local intensity can even be enhanced compared to the intensity of the incident light. This is often the case for metal-coated tips, where the free electrons in the metal enhance the lightning-rod effect and in some cases provide additional enhancement of the local intensity due to surface plasmon resonances [23]. The tip thus acts as an antenna that further focuses the (diffraction limited) excitation beam toward nanoscale dimensions of typically 20–30 nm. The degree of localization of the excitation light now depends solely on the shape and size of the tip and (in first approximation) not on the wavelength of light anymore.

The localized fields around the tip apex interact with the sample surface, which modifies the back-scattered field in amplitude and phase, depending on the local optical sample properties. By recording the back-scattered light as a function of tip position, nanoscale-resolved images of the sample's optical properties are obtained. The strongly enhanced and localized intensity around a metalized tip is widely used for probing vibrational excitations such as the ones of molecules or phonons, where vibrational spectra provide information on the chemical composition of the underlying material.

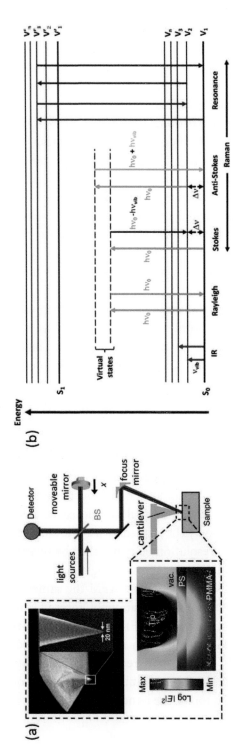

FIGURE 5.9   Aperture-less NSOM. (a) A sharp tip or cantilever is actively kept in close proximity to a sample using an atomic force microscope. The tip of the cantilever is illuminated by a light source that generates a strong local intensity around the tip due to the "lightning-rod effect", which is used as an excitation source. The signal scattered or emitted by the sample is passed through a Michelson interferometer for interferometric detection. The inset shows a numerical simulation of the local intensity near the tip of the cantilever, together with a scanning electron microscopy image of a typical cantilever used in aperture-less NSOM. (b) Aperture-less NSOM is most often used for chemical mapping of materials by probing vibrational energy levels. Vibrational spectroscopy can be performed in different geometries, indicated in the Jablonski diagram and further explained in the text. ([a] Adapted with permission under a Creative Commons CC-BY 4.0 license from [27]. Copyright 2020 The Authors. The SEM image is reprinted with permission from [28]. Copyright 2017 American Chemical Society. [b] Reprinted with permission under a Creative Commons CC-BY 4.0 license from [29]. Copyright 2021 The Authors.)

The different modes in which an s-SNOM is typically operated are shown in Figure 5.9b. Atoms and molecules in a sample are excited by the enhanced intensity around the tip region, which firstly results in Rayleigh scattering of the sample. Rayleigh scattering concerns the elastically scattered light by particles much smaller than the wavelength of the incident light, whose intensity is inversely proportional to the fourth power of the wavelength [30]. Rayleigh scattering itself does not directly contain information on the chemical properties of the underlying material; however, part of the scattered light will have undergone a change in energy (i.e., it is scattered inelastically) due to the presence of vibrational modes in the material. This results in a very weak signal at energies above and below the incident light energy (i.e., at wavelengths below and above the excitation wavelength) due to Raman scattering.

Like Rayleigh scattering, the intensity of Raman scattered light is roughly proportional to the square of the excitation intensity [31]. This provides a strong argument to use metallized tips for Raman imaging due to the local intensity enhancement around the tip apex. Nevertheless, tip-enhanced Raman scattering (TERS) also exhibits weak signals, and it is usually necessary to separate the Raman scattered light from the Rayleigh signal and reflected laser signal using high-quality optical filters. An example of a TERS experiment is shown in Figure 5.10a, where a silver-coated tip excited with a 634 nm laser is used to image a porphyrin molecule immobilized on a copper substrate [32]. The silver tip in this case is atomically sharp, while the presence of a second metal interface (the copper substrate) further enhances the local excitation intensity by so-called cavity plasmons. This enabled the collection of Raman spectra with high signal-to-noise ratio, as shown in the middle panel that clearly reveal different vibrational modes as indicated by the shaded areas. The atomically sharp silver tip resulted in a spatial resolution of <1 nm thereby revealing the spatial distribution of vibrational modes across a single molecule.

Because the s-SNOM essentially operates on an AFM the same measurement also enables the simultaneous application of forces. This is shown in Figure 5.10b that depicts correlative TERS and force spectroscopy of a single-walled carbon nanotube. When the nanotube is compressed, a new vibrational mode (depicted in red in Figure 5.10b(A)) appears due to local deformations of the nanotube. This mode is only detected when the tip is directly on top of the nanotube, confirming it is due to compression forces.

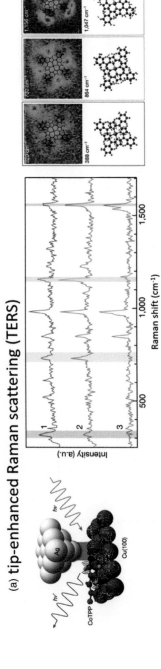

FIGURE 5.10 Examples of apertureless NSOM in materials science. (a) Tip-enhanced Raman scattering (TERS) of the porphyrin CoTPP immobilized on a copper substrate at 6 Kelvin. Using an atomically sharp silver tip the excitation volume is smaller than the molecule itself, enabling Raman spectroscopy at the atomic level. The middle pane shows the spectra recorded on the cobalt (1), pyrrole (2), and phenyl groups (3). The right panel shows spatial maps of the vibrational modes at different Raman shifts, enabling the direct visualization of the location of the vibrational mode. The bottom illustration shows the vibrational mode that is being imaged on the porphyrin molecule.   (*Continued*)

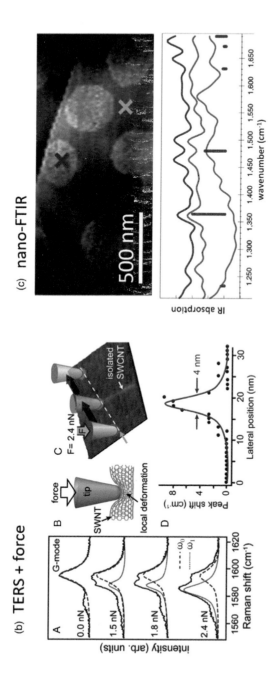

FIGURE 5.10 (CONTINUED) Examples of apertureless NSOM in materials science. (b) Correlative TERS and force spectroscopy of a single-walled carbon nanotube. When the nanotube is compressed, a new vibrational mode (depicted in red in b(A)) appears due to local deformations of the nanotube. This mode is only detected when the tip is directly on top of the nanotube, confirming it is due to compression forces. (c) Nanoscale infrared absorption spectroscopy (nano-FTIR) of a monolayer photopolymer on a passivated graphite surface. The top panel shows a standard height image collected using AFM, where the crosses indicate the locations at which the infrared absorption spectra in the bottom panel are recorded. The spectra clearly reveal infrared absorption by vibrational modes in the photo-polymer, which are not present when the substrate itself is imaged. ([a] Adapted with permission from [32]. Copyright 2019 Springer Nature. [b] Reproduced with permission from [34]. Copyright 2009 Springer Nature. [c] Adapted with permission from [35]. Copyright 2021 Springer Nature.)

Instead of imaging the Raman scattered signal, vibrational fingerprints can also be obtained by direct infrared absorption measurements. Herein the tip is illuminated by a broadband infrared beam that locally illuminates the sample with a wide spectrum rather than a narrow-band laser line. The backscattered light from the tip is detected, which is again enhanced due to the metallized tip as described above. Often the backscattered light is detected interferometrically using a Michelson interferometer as depicted in Figure 5.9a. The sample stage is placed into one arm of a conventional Michelson interferometer, while a mirror on a piezo stage is placed into another reference arm. Recording the backscattered signal while translating the reference mirror yields an interferogram, while the subsequent Fourier transform of this interferogram returns the near-field spectrum of the sample [33]. This approach is called Fourier transform infrared nanospectroscopy (nano-FTIR).

An example of a nano-FTIR experiment on a polymer film is shown in Figure 5.10c. The image shows the infrared absorption spectra recorded from a monolayer photopolymer on a passivated graphite surface. The crosses indicate the locations at which the infrared absorption spectra in the bottom panel are recorded. The spectra clearly reveal infrared absorption by vibrational modes in the photopolymer, which are not present when the substrate itself is imaged.

Nano-FTIR has become a popular approach to quantify vibrational modes in thin structures such as polymer films and 1D materials such as graphene [36]. Tip-enhanced imaging in the infrared has a clear benefit over diffraction-limited far-field imaging because of the long wavelengths used: while diffraction-limited resolution deteriorates for increasing wavelengths, the spatial resolution in nano-FTIR is largely determined by the tip-dimensions. This provides a broadband >100-fold increase in spatial resolution across infrared wavelengths over diffraction-limited optics. In the future, the use of correlative near-field imaging will likely provide a new level of understanding by correlating nanoscale vibrational fingerprints with nanoscale topography and nano-mechanical properties [17].

All-in-all near-field microscopy is a very powerful tool bypassing the diffraction limit without the need of special dyes, sample preparation, or complex analysis. However, its use is nowadays still limited by the difficulty of the instrumentation needed and is not yet commonly used to image biological or complex synthetic architectures.

REFERENCES

1. S. Geissbuehler, C. Dellagiacoma, and T. Lasser, "Comparison between SOFI and STORM," *Biomed. Opt. Express* **2**, 408 (2011).

2. T. Dertinger, R. Colyer, G. Iyer, S. Weiss, and J. Enderlein, "Fast, background-free, 3D super-resolution optical fluctuation imaging (SOFI)," *Proc. Natl. Acad. Sci.* **106**, 22287–22292 (2009).

3. J. M. Mendel, "Tutorial on higher-order statistics (spectra) in signal processing and system theory: theoretical results and some applications," *Proc. IEEE* **79**, 278–305 (1991).

4. T. Dertinger, R. Colyer, R. Vogel, J. Enderlein, and S. Weiss, "Achieving increased resolution and more pixels with Superresolution Optical Fluctuation Imaging (SOFI)," *Opt. Express* **18**, 18875 (2010).

5. B. Moeyaert, W. Vandenberg, and P. Dedecker, "SOFIevaluator: a strategy for the quantitative quality assessment of SOFI data," *Biomed. Opt. Express* **11**, 636 (2020).

6. A. Alva, E. Brito-Alarcón, A. Linares, E. Torres-García, H. O. Hernández, R. Pinto-Cámara, D. Martínez, P. Hernández-Herrera, R. D'Antuono, C. Wood, and A. Guerrero, "Fluorescence fluctuation-based super-resolution microscopy: basic concepts for an easy start," *J. Microsc.* **288**, 218–241 (2022).

7. N. Gustafsson, S. Culley, G. Ashdown, D. M. Owen, P. M. Pereira, and R. Henriques, "Fast live-cell conventional fluorophore nanoscopy with ImageJ through super-resolution radial fluctuations," *Nat. Commun.* **7**, 12471 (2016).

8. Y. Han, X. Lu, Z. Zhang, W. Liu, Y. Chen, X. Liu, X. Hao, and C. Kuang, "Ultra-fast, universal super-resolution radial fluctuations (SRRF) algorithm for live-cell super-resolution microscopy," *Opt. Express* **27**, 38337 (2019).

9. M. Browne, H. Gribben, M. Catney, C. Coates, G. Wilde, R. Henriques, J. T. Cooper, and A. Mullan, "Real time multi-modal super-resolution microscopy through Super-Resolution Radial Fluctuations (SRRF-Stream)," in *Single Molecule Spectroscopy and Superresolution Imaging XII*, I. Gregor, Z. K. Gryczynski, and F. Koberling, eds. (SPIE, 2019), p. 42.

10. A. Egner, C. Geisler, and R. Siegmund, "STED nanoscopy," in *Nanoscale Photonic Imaging*, T. Salditt, A. Egner, and D. R. Luke, eds., Topics in Applied Physics (Springer International Publishing, 2020), Vol. 134, pp. 3–34.

11. C. J. R. Sheppard, "Super-resolution in confocal imaging," *Optik* **80**, 53–54 (1988).

12. C. B. Müller and J. Enderlein, "Image scanning microscopy," *Phys. Rev. Lett.* **104**, 198101 (2010).

13. J. Huff, "The airyscan detector from ZEISS: confocal imaging with improved signal-to-noise ratio and super-resolution," *Nat. Methods* **12**, i–ii (2015).

14. K. Korobchevskaya, B. Lagerholm, H. Colin-York, and M. Fritzsche, "Exploring the potential of airyscan microscopy for live cell imaging," *Photonics* **4**, 41 (2017).

15. I. Gregor and J. Enderlein, "Image scanning microscopy," *Curr. Opin. Chem. Biol.* **51**, 74–83 (2019).

16. G. M. R. De Luca, R. M. P. Breedijk, R. A. J. Brandt, C. H. C. Zeelenberg, B. E. De Jong, W. Timmermans, L. N. Azar, R. A. Hoebe, S. Stallinga, and E. M. M. Manders, "Re-scan confocal microscopy: scanning twice for better resolution," *Biomed. Opt. Express* **4**, 2644 (2013).

17. K. Bian, C. Gerber, A. J. Heinrich, D. J. Müller, S. Scheuring, and Y. Jiang, "Scanning probe microscopy," *Nat. Rev. Methods Primer* **1**, 36 (2021).

18. G. Binnig, H. Rohrer, C. Gerber, and E. Weibel, "Surface studies by scanning tunneling microscopy," *Phys. Rev. Lett.* **49**, 57–61 (1982).

19. G. Binnig, C. Quate, and C. Gerber, "Atomic force microscope," *Phys. Rev. Lett.* **56**, 930–933 (1986).

20. E. H. Synge, "XXXVIII. A suggested method for extending microscopic resolution into the ultra-microscopic region," *Lond. Edinb. Dublin Philos. Mag. J. Sci.* **6**, 356–362 (1928).

21. D. W. Pohl, W. Denk, and M. Lanz, "Optical stethoscopy: image recording with resolution λ/20," *Appl. Phys. Lett.* **44**, 651–653 (1984).

22. A. Lewis, M. Isaacson, A. Harootunian, and A. Muray, "Development of a 500 Å spatial resolution light microscope," *Ultramicroscopy* **13**, 227–231 (1984).

23. H. Lee, D. Y. Lee, M. G. Kang, Y. Koo, T. Kim, and K.-D. Park, "Tip-enhanced photoluminescence nano-spectroscopy and nano-imaging," *Nanophotonics* **9**, 3089–3110 (2020).

24. P. Hinterdorfer, M. F. Garcia-Parajo, Y. F. Dufrene, and Y. F. Dufrêne, "Single-molecule imaging of cell surfaces using near-field nanoscopy," *Acc. Chem. Res.* **45**, 327–336 (2012).

25. T. S. Van Zanten, J. Gómez, C. Manzo, A. Cambi, J. Buceta, R. Reigada, and M. F. Garcia-Parajo, "Direct mapping of nanoscale compositional connectivity on intact cell membranes," *Proc. Natl. Acad. Sci.* **107**, 15437–15442 (2010).

26. J. T. Krug, E. J. Sánchez, and X. S. Xie, "Design of near-field optical probes with optimal field enhancement by finite difference time domain electromagnetic simulation," *J. Chem. Phys.* **116**, 10895–10901 (2002).

27. L. Mester, A. A. Govyadinov, S. Chen, M. Goikoetxea, and R. Hillenbrand, "Subsurface chemical nanoidentification by nano-FTIR spectroscopy," *Nat. Commun.* **11**, 3359 (2020).

28. P. Verma, "Tip-enhanced Raman spectroscopy: technique and recent advances," *Chem. Rev.* **117**, 6447–6466 (2017).

29. A. Orlando, F. Franceschini, C. Muscas, S. Pidkova, M. Bartoli, M. Rovere, and A. Tagliaferro, "A comprehensive review on raman spectroscopy applications," *Chemosensors* **9**, 262 (2021).

30. C. F. Bohren and D. R. Huffman, *Absorption and Scattering of Light by Small Particles* (Wiley-VCH, 1983).

31. L. Novotny and B. Hecht, *Principles of Nano-Optics* (2009).

32. J. Lee, K. T. Crampton, N. Tallarida, and V. A. Apkarian, "Visualizing vibrational normal modes of a single molecule with atomically confined light," *Nature* **568**, 78–82 (2019).

33. F. Huth, A. Govyadinov, S. Amarie, W. Nuansing, F. Keilmann, and R. Hillenbrand, "Nano-FTIR absorption spectroscopy of molecular fingerprints at 20 nm spatial resolution," *Nano Lett.* **12**, 3973–3978 (2012).

34. T. Yano, P. Verma, Y. Saito, T. Ichimura, and S. Kawata, "Pressure-assisted tip-enhanced Raman imaging at a resolution of a few nanometres," *Nat. Photonics* **3**, 473–477 (2009).

35. L. Grossmann, B. T. King, S. Reichlmaier, N. Hartmann, J. Rosen, W. M. Heckl, J. Björk, and M. Lackinger, "On-surface photopolymerization of two-dimensional polymers ordered on the mesoscale," *Nat. Chem.* **13**, 730–736 (2021).

36. X. Chen, D. Hu, R. Mescall, G. You, D. N. Basov, Q. Dai, and M. Liu, "Modern scattering-type scanning near-field optical microscopy for advanced material research," *Adv. Mater.* **31**, 1804774 (2019).

CHAPTER 6

Quantitative Analysis for Single-Molecule Localization Microscopy

"From PSF to Information"

Teun A.P.M. Huijben, Rodolphe Marie, and Kim I. Mortensen

Technical University of Denmark (DTU), Kongens Lyngby, Denmark

Single-molecule localization microscopy (SMLM) revolutionized the field of microscopy by allowing imaging below the diffraction limit. For decades, the resolution of microscopy has been limited by Abbe's diffraction limit [1]. This limit dictates that two emitters cannot be resolved when they are less than $\lambda/2NA$ apart, with λ the emission wavelength of the emitter and NA the numerical aperture of the objective. For an average microscope operating in the visible light spectrum, this translates to an approximate distance of 200 nm. Single-molecule localization microscopy as described in Chapter 2 shared the 2014 Noble Prize in Chemistry for super-resolution microscopy together with stimulated emission depletion (STED) microscopy. SMLM overcomes the diffraction limit by employing the clever trick of letting emitters repeatedly switch on and off over time [2]. When imaging these blinking emitters for many frames, in each frame only a sparse subset of the fluorophores is emitting light, resulting in isolated diffraction-limited spots in the image. The location of every individual spot can be determined with nanometer precision by fitting a model of the point spread function (PSF). The final super-resolution image

DOI: 10.1201/9781003220688-6

is a rendering of the coordinates of all calculated positions of the spots in all frames.

In this chapter, we will delve into the topic of how information is extracted from a spot of light in a microscopy image. We start by describing the characteristics of the PSF; we explain how to model it and present commonly used fitting strategies. Then we will cover various strategies that allow determining the position and orientation of the emitter with greater precision.

6.1 FITTING THE POINT SPREAD FUNCTION

Light emitted by an emitter travels through the microscope and forms an image that is recorded by a camera. In case of a point-source, the spot of light detected on the camera is not infinitely small because the light interacts with the optical components in the microscope. Especially, diffraction [1] of the light at the objective of the microscope is the reason the detected spot of light has a finite width and consequently appears as a blurry spot. The shape of the intensity distribution in that spot is commonly referred to as the point spread function (PSF). Specifically, it is defined as the probability distribution for a photon to arrive at a certain location on the camera. The PSF is dictated by the physical parameters of the microscope and the light emitter. To translate the theoretical probability distribution of the PSF to a typical pixelated microscopy image, one needs to integrate this probability distribution over the area of the pixel, multiply it with the total number of photons that are collected from the point-source, and add contributions of background photons and noise. In order to extract information from the blurry spot, an analytical parametric form of the PSF is required such that it can be fitted to the experimentally acquired spot in the image. In the following section we will describe such PSF models.

6.1.1 Model for the PSF
6.1.1.1 The Airy PSF
Due to the finite size of the circular back aperture of the objective lens, light cannot be focused onto an infinitely small spot. As discussed in Chapters 1 and 2 the diffraction-limited spot, the PSF, therefore has a finite width. Over the last decades, multiple models for the PSF have been derived and used to fit the measured spots.

The textbook example of the PSF is the Airy distribution (Figure 6.1a), which is rotationally symmetric and consists of a dominant central peak surrounded by diffraction rings that decrease in intensity following $1/r^3$,

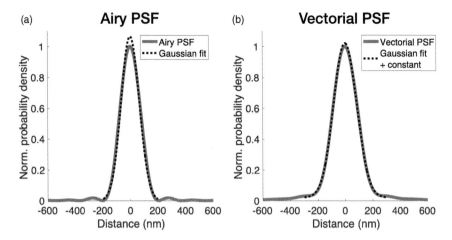

FIGURE 6.1   Airy and vectorial PSFs. (a) The Airy PSF model with a dominant central peak and diffraction rings. The 2D-Gaussian is fitted to the part of the PSF up until its first zero at ±200 nm. (b) The vectorial PSF model of [4] with a dominant central peak and characteristic shoulders. The 2D Gaussian plus constant is fitted to the part of the PSF up until the end of the shoulders at ±300 nm.

where r is the distance from the center. The Airy distribution follows from the diffraction of a plane wave from a circular aperture that represents the shape of the lens [1]. The light entering the circular objective can be considered a plane wave when it originates from an isotropic point-source located far away. Often, the Airy PSF is approximated by a 2-dimensional Gaussian distribution (Figure 6.1a) for the following three reasons: the low-intensity rings in the Airy distribution do not contain many photons and often disappear in the noise level, the central peak resembles a Gaussian-like shape, and a Gaussian distribution is analytically and computationally cheaper to fit [3].

6.1.1.2 The Vectorial PSF

The Airy PSF only holds for a point-source of light that is located far away from the optical system and emitting light isotropically, i.e., it emits uniformly in all directions. Fluorescent molecules, however, are not isotropic sources of light, but emit via a dipole transition [5]. Fluorescent molecules in general contain a chemical group that has a transition dipole moment, which can intuitively be seen as the vector along which electrons oscillate after being excited by an electromagnetic wave. When excited, the fluorophore goes into the excited state, with a probability proportional to the overlap of its excitation transition dipole moment with the polarization vector of the excitation light. When returning from the excited state to the ground state, the molecule emits light in a doughnut-shaped emission pattern with maximum emission intensity perpendicular to the emission transition dipole moment (which is for many common fluorophores nearly parallel to the excitation dipole moment). For this reason, the light emitted by a fluorophore follows a dipolar emission pattern (Figure 6.2a, b).

In the dipolar emission pattern, the power is emitted away from the fluorophore in a direction along the plane perpendicular to the dipole transition moment, and no power is emitted along the direction of the dipole transition moment itself. Since this emission pattern is far from isotropic, the Airy PSF model is not valid anymore. Based on the dipolar emission of a fluorophore, a vectorial PSF of a fluorophore with fixed orientation can be derived analytically using the vector diffraction approach [4, 6], based on earlier work of dipole emitters near planar surfaces [7–9]. The shape of the PSF is no longer rotationally symmetric and depends on the orientation of the fluorophore's dipole moment with respect to the glass coverslip

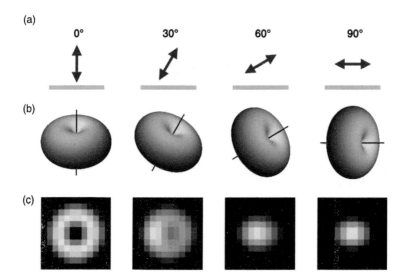

FIGURE 6.2 PSF of fluorophore with fixed orientation. (a) Schematic representation of four orientations of the transition dipole moment (red) with respect to the glass coverslip (blue), where the mentioned angles represent the out-of-plane angles measured away from the optical axis. (b) The characteristic 'donut-shaped' dipole emission patterns for the four dipole moments in (a). (c) The PSFs corresponding to the orientations of the dipole moments in (a), for a wavelength of 500 nm, a microscope numerical aperture of 1.49, and a pixel size of 40 nm. (Calculated with the script from [4].)

(Figure 6.2c). One typically assumes that the fluorophore is located in water and separated from the objective by a glass coverslip. The analytical expression for the vectorial PSF can be fitted to experimental images to extract the fluorophore position and orientation with optimal precision (see below) [4, 6]. One could argue that the PSF for a fixed dipole is well-approximated by a Gaussian with two different variances, provided that the dipole is oriented parallel to the glass surface (right-most image in Figure 6.2c), and therefore propose to fit the spot with a Gaussian. However, fitting a Gaussian to these spots results in large biases, on the order of tens of nanometers, demonstrating the necessity for the vectorial PSF [6, 8, 10–12].

In most experimental settings, the fluorophore will not be fixed in orientation, it is either diffusing (translating and rotating) in solution or bound to a site of interest via a long and/or flexible linker. In the particular case when the fluorophore's rotational diffusion is fast compared to its excited-state lifetime, it essentially emits isotropically. The electromagnetic fields of the freely rotating dipole are therefore obtained by considering the dipolar emission pattern of a fixed dipole averaged over all possible orientations, each with equal probability. The shape of the resulting PSF is rotationally symmetric and, compared to the Airy PSF, lacks the diffraction rings because the emission wavefronts do not form a plane wave. The PSF is then characterized by a shoulder of nearly constant intensity around the central peak (Figure 6.1b) [4, 6].

The analytical expression for the rotationally free vectorial PSF is elaborate and expensive to fit. However, a 2-dimensional Gaussian plus constant approximates the rotationally free vectorial PSF very well when the spot is cropped within the shoulders (Figure 6.1b). A maximum-likelihood fit of a Gaussian plus constant is therefore a very efficient method for fitting spots, and it has been mathematically and experimentally proven that this method achieves the best possible precision on positional estimates of the fluorophore [4]. Note that when fitting a Gaussian plus constant to the central peak of the spot, the number of photons is underestimated to around 60% of the true value, because the source photons that are present in the shoulders and the tails of the distribution are considered as "background". This, however, does not sacrifice precision about the emitter's position since the shoulders and the slowly varying background provide little information to this end.

6.1.2 Fitting

In the previous section, we have discussed multiple PSF models (Airy, Gaussian, and vectorial) that can be used to fit the spot and thereby extract information about the position (and orientation) of molecules. However, researchers use various fitting procedures to perform fits, and it is vital to understand the differences between procedures and their effect on the results.

The aim of fitting a model to data is to extract information from the data. The quality of the fit is expressed in terms of precision and accuracy, where the precision describes the spread of the obtained values around their mean and the accuracy describes the deviation of the mean of the obtained values from the true position [13]. The precision is usually expressed as the standard deviation of repeated measurements, and the accuracy as the bias between the average obtained value and the true value. When fitting a PSF model to a spot, ideally, we want the bias to be zero and the precision to be as high as possible. The best theoretically achievable precision for an unbiased estimator is given by the Cramér-Rao lower bound (CRLB) [14, 15]. For SMLM, the localization precision typically scales with $1/\sqrt{N}$, with N the number of collected photons and can therefore be improved by collecting more photons. This CRLB can be used to calculate the best possible precision for a certain estimator. In turn, it may be used to verify that the experimentally obtained precision does not violate information theory, because a violation of the information limit is only possible when the fit is biased. We will describe the two most commonly used fitting routines, least-squares and maximum-likelihood estimation [16].

6.1.2.1 Least-Squares

In unweighted least-squares fitting (LS), the parameters of the PSF model are varied in order to minimize the mismatch between the spot and the PSF model. The mismatch is quantified by taking the squared difference between the data and the model summed over all pixels, clarifying the name least-squares. LS assigns equal weights to all pixels, which is suboptimal because photon counts in a pixel follow a Poisson distribution. Hence, the variance is equal to the mean, i.e., the expected number of photons in the respective pixel. Therefore, pixels with a high number of photons have a larger variance and should be given a lower weight in the fit. Weighted least-squares (WLS) solves this issue by assigning each pixel a weight according to either the actual or expected number of photons in the pixel (depending on the implementation), making sure the pixels with a larger variance are given a lower weight. In this way, the weighting gives more importance to the tails of the PSF. Even though weighting the pixels is necessary, this results in two problems: a low photon count in the tails (due to Poisson statistics) gives an artificially high weight to the pixels with low/zero photons, and the PSF model has to correctly describe the tails and cannot be an approximation (like the Airy and Gaussian PSFs). The alternative fitting procedure eliminating all these problems is maximum-likelihood estimation (MLE).

6.1.2.2 Maximum-Likelihood Estimation

MLE requires a model for the PSF, as well as a model for the noise distribution for the number of photons in a pixel. Using these two models, a likelihood function is constructed that describes the probability that the observed data, the PSF, is generated by the suggested models. The parameters of interest are found by maximizing the likelihood function. A key advantage of MLE is that it is theoretically proven to be the optimal fitting procedure. This means that it achieves the best-possible precision, given by the CRLB when the estimator is unbiased and the image is formed by a large total number of photons [14, 15].

The difference between MLE and (W)LS is that the latter implicitly assumes a Gaussian distribution for the noise, which is neither the case for photon statistics in general nor in the presence of the commonly used sCMOS and EMCCD cameras. For these cameras, the gain, multiplication noise, and readout noise have to be calibrated carefully (ideally per pixel [17]) in order for MLE to obtain the CRLB. LS and WLS have the advantage that knowing the noise distribution is not necessary, whereas LS not even requires the variance, which makes it an easy tool for when the noise is not characterized [16]. For high photons counts, WLS performs equally well as MLE, although LS squanders a substantial amount of information because of its failure to account correctly for the weights. However, at lower photon counts, MLE will significantly outperform (W)LS [3, 4, 18]. Note that all the above-mentioned considerations assume the PSF model is correct, a wrong PSF model will lead to a bias or a lower precision.

6.2 PSF ENGINEERING – MAKING THE PSF MORE INFORMATIVE

In the previous section, we have seen how the PSF encodes positional and orientational information of the fluorophore. This information is extracted from a spot by fitting the appropriate PSF model. Even though the position and orientation can be obtained with high precision and accuracy [4, 12, 19], these methods have their limitations. First of all, it is challenging to get information about the 3D position of the fluorophore, since the PSF only varies slowly with the axial position [20]. Secondly, it is difficult to determine the out-of-plane (polar) angle of oriented fluorophores from in-focus PSFs, since the PSF shape does not vary much as a function of the out-of-plane angle when the fluorophore is oriented perpendicular to the optical axis. Furthermore, the estimates are heavily influenced by a slight defocus, or by a lower degree of orientational constraint ('wobbling') of the fluorophore [21]. In this section, we will outline various PSF engineering strategies, which give the PSF a more informative shape as a function of the axial position and orientation of the fluorophore.

6.2.1 3D Localization

The PSF is not informative about the axial position of the fluorophore, since the PSF shape and intensity only vary slowly with axial position, and the PSF is symmetric for positions above or below the focal plane. Here, we will discuss various approaches to encode the axial position more informatively into the shape of the PSF. This with the goal that the axial position of the fluorophore can be deducted from the shape of the spot.

6.2.1.1 Imaging Multiple Planes

Given the reasons above, it is hard to estimate the 3D position of a fluorophore from a 2D slice of the PSF [20]. One solution is to sample the PSF at different z-positions and perform a combined fit to the set of images. The simplest implementation is bi-focal imaging [22, 23] where the PSF is sampled at two positions: at the focal plane and at a distance away from the focal plane (either 500 nm or 1 μm defocus, depending on the implementation). The combination of the two images allows fitting of the full 3D PSF model and thereby extraction the 3D position of the fluorophore. The method is not limited to two but can be extended to multiple planes [24, 25] to obtain a better axial precision.

6.2.1.2 Phase Manipulation (PSF Engineering)

An alternative approach to encode the fluorophore's axial position into the PSF shape is by deliberately altering the phase of the emitted light. In this way, one specifically 'designs' the microscopy image to contain additional information, commonly referred to as 'PSF engineering'. One of the earliest ideas was to add a cylindrical lens to the optical path of the microscope to generate astigmatism, which is a type of aberration where the ellipticity of the PSF depends on the axial position of the fluorophore [26–28].

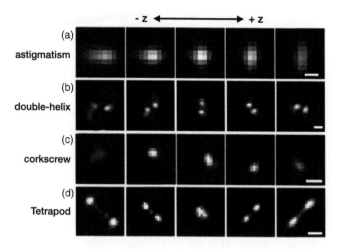

FIGURE 6.3   Various experimentally measured engineered PSFs encoding the 3D position of the fluorophore. (a) Astigmatism. Scale bar is 500 nm. (b) Double-helix PSF. Scale bar 2 μm. (c) Corkscrew PSF. Scale bar 1 μm. (d) Tetrapod PSF. Scale bar 1 μm. (Adapted with permission from [20]. Copyright 2017 American Chemical Society.)

The orientation and aspect ratio of the elliptical PSF are found via fitting and encode information about the axial position (Figure 6.3a).

A more sophisticated way to change the phase is positioning a phase plate or spatial light modulator (SLM) at the back focal plane of the microscope. The back focal plane is conjugated to the focal plane and represents the Fourier transform (i.e., frequency information) of the image. Placing a phase-altering component at this location allows one to modify the emission beam phase profile. Such elements allow one, in theory, to apply any phase alteration to the light. Many different phase alterations that add axial information to the shape of the PSF have been published. Examples are the double-helix, corkscrew, and Tetrapod PSFs (Figure 6.3b–d). The double-helix PSF contains two spots of light that revolve around each other [29–33], the corkscrew PSF is a single spot that rotates around the true x, y-position [34], and the Tetrapod PSF (also called the saddle-point PSF) has two strong lobes that vary in distance and angle [35, 36], all as function of the axial position of the fluorophore.

6.2.2 Fluorophore Orientation

The information that can be encoded in the PSF is not limited to the axial position. There is also a large interest in deciphering the orientation of a fluorescent molecule from the PSF to study the orientational structure of materials. This can entail the 2D or 3D orientation of the transition dipole moment of the fluorophore and its level of wobbling. Here we will discuss various approaches to encode the orientation and wobbling of a fluorophore more informatively into the shape of the PSF. Where some methods are specifically designed to retrieve the orientation, and thereby lose the ability to obtain the position precisely, others make efforts to obtain both.

6.2.2.1 Defocus

One of the first methods that increased the sensitivity to the orientation of the fluorophore was defocusing, based on the observation that introducing a defocus aberration enlarges the shape-variation of the PSF with respect to the orientation (Figure 6.4a). After deliberately positioning the fluorophore out of focus, similar to the multiplane methods described above, the obtained image can be fitted to deduce the fluorophore's orientation [37, 38], where the level of defocus is usually set to 1 μm [39–41]. Since defocusing lowers the accuracy and precision on the estimation of the position of the fluorophore, efforts are made to obtain the position along with the orientation by imaging both in- and out-of-focus. The position is calculated from the in-focus image and the orientation from the out-of-focus one [42]. This, however, has the drawback that only a subset of the total number of photons is used for either purpose.

6.2.2.2 Back Focal Plane Imaging

The method of determining the orientation by using defocus relies on the ability to determine the perfect focus, which is difficult in practice.

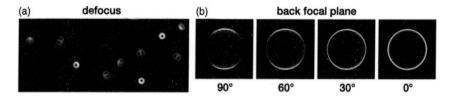

FIGURE 6.4 Approaches to retrieve the orientation of the fluorophore. (a) Defocus imaging enhances the orientation contrast of PSF. Adapted with permission from [39]. Copyright 2009 Optical Society of America. (b) Simulated back focal plane (BFP) images for four different out-of-plane angles, where 90° means that the transition dipole moment is parallel to the glass coverslip and 0° means that it is perpendicular to the same surface. (Adapted with permission from [46]. Copyright 2013 John Wiley and Sons.)

Another technique, not sensitive to the level of defocus, positions the camera at the back focal plane (BFP) of the microscope, i.e., not at the image plane. The intensity distribution in the BFP represents the angular information of the emitted light and is therefore highly affected by the orientation of the dipole emitter (Figure 6.4b). By comparing the intensity pattern of the obtained BFP image, one can obtain information about the orientation of the fluorophore [43]. Other work extends this technique by manipulating the BFP information using a phase mask or SLM to enhance the orientational contrast [44, 45]. A disadvantage of this method is that all positional information is lost in the BFP, making it less suitable for super-resolution microscopy applications. Additionally, the BFP contains an overlap of the angular information from all fluorophores in the field of view. To overcome this problem, one has to employ point-scanning methods and access the orientation of all single molecules one-by-one.

6.2.2.3 Polarization

As explained above, the response of a fluorophore to the incident light depends on the angle between the fluorophore's excitation transition dipole moment and the polarization of the incident light. The probability of absorbing a photon scales with the dot product between these two vectors. Knowing this, the polarization of both the incident and emitted light contains information about the orientation of the fluorophore, and both can be utilized to retrieve this information.

The simplest option is to control the polarization of the incident light. When illuminating the sample with two different polarizations (along the x- and y-direction), the corresponding intensities per polarization direction are calculated and the ratio between the two indicates the in-plane (x, y) orientation of the fluorophore [47].

Along the same line of reasoning, the polarization of the fluorophore's emitted light contains the same information. A commonly used technique is to split the collected fluorescence into orthogonal polarizations and infer the fluorophore's in-plane orientation from the ratio between the channels. Implementations include strategies where the polarization is split in either two [48] or four [49–51] channels, which allows finding the in-plane angle and the degree of wobbling. A more elaborate implementation splits both the excitation in four channels and the emission in two channels, allowing the recovery of the full 3D orientation and wobbling of the fluorophore [52, 53].

The polarization-splitting approaches have multiple disadvantages, despite providing additional information about the fluorophore's orientation. First of all, the total number of photons are divided over different channels reducing the intensity of the PSF. Secondly, one either needs multiple cameras to record every channel, or to project each channel onto a different area of the same camera, thereby reducing the field of view.

6.2.2.4 Phase Manipulation (PSF Engineering)

In line with the demonstrated PSF engineering solutions that encode the axial position in the PSF, similar methods have been developed that encode the fluorophore's orientation.

Multiple different PSFs are published that add orientational information to the shape of the PSF: of which the double-helix PSF, quadrated PSF, bisected PSF, Tri-spot PSF, and vortex PSF are some examples. The double-helix PSF was originally applied for providing additional axial information, as mentioned above. However, information about the molecular orientation is reflected in the asymmetry between the two lobes of the double-helix PSF [29]. The quadrated, bisected, and Tri-spot PSFs, are all generated by a combination of a phase mask and polarization splitting into two channels. The PSF has the shape of a four-pointed star [54], two parallel lobes [55], or three lobes in the shape of a triangle [56], respectively, where the relative intensities, distances and/or angles depend on the orientation of the fluorophore. In contrast to the quadrated PSF that only encodes orientation, the bisected PSF also contains positional information and the Tri-spot PSF even contains wobbling.

Recently, two versions of the vortex PSF are published that both encode the 3D position, 3D orientation, and wobbling of the fluorophore. The implementation of Hulleman et al. [21] only requires a vortex phase plate at the BFP, which can easily be added to most microscopes. The version of Ding et al. [57] uses the vortex phase plate in combination with polarization splitting, which requires a more elaborate optical setup, but has a slightly better orientational sensitivity.

6.3 MODULATION ENHANCED LOCALIZATION MICROSCOPY

Ever since the invention of super-resolution localization microscopy, the field has aimed to improve the resolution by bringing it down to the nanometer length scale. Where localization microscopy typically delivers a resolution of 20 nm, values close to a single nanometer are required to obtain information about underlying interactions at the molecular scale. To achieve a better resolution, one has to think about reducing the size of the linker between the fluorophore and the object of interest, overcoming the problem of limited labeling efficiency by data fusion [58, 59] and improving the localization precision. Especially the latter is typically addressed, since the localization precision directly depends on the number of photons and scales with $1/\sqrt{N}$, allowing a straightforward way to improve the localization precision by *just* collecting more photons. However, this limits the time resolution and the total time the emitter can be imaged for. Efforts have been made to increase the number of photons by engineering brighter fluorophores [60], avoiding bleaching by imaging at cryogenic temperatures [61], improving drift correction methods that allow for longer imaging, and employing fluorescent enhancement by placing plasmonic metal structures close to the fluorophore [62].

All these approaches rely on the principle that every collected photon contains information about the fluorophore's position, and the collection of more and more photons improves the statistical precision of the estimated position. Initialized by the emergence of the minimal emission fluxes (MINFLUX) concept, this paradigm has shifted recently by modulation enhanced localization microscopy techniques.

6.3.1 MINFLUX

The idea behind MINFLUX is to localize a single fluorophore by moving a doughnut-shaped *illumination* spot over the sample to find the position where the fluorophore is in the illumination minimum of the doughnut, the valley [63]. The doughnut spot is moved to four positions around a region of interest: three positions in a triangular shape of size L and one in the middle (Figure 6.5). Since the illumination pattern of the doughnut is known, the fluorophore position can be calculated by triangulating the photon counts obtained at the four visited positions. The localization precision scales with $L/\sqrt{N}$, which is favourable since it can be improved by choosing a small triangulation length L. In general, one starts with a larger triangulation and iteratively decreases L to ensure a high localization precision.

The intuitive rationale of why MINFLUX can improve the localization precision by more than an order of magnitude is that extra information about the illumination pattern allows for extracting extra information from the sample. Instead of illuminating the entire field of view (FOV) homogeneously and retrieving the fluorophore's position as the centroid of

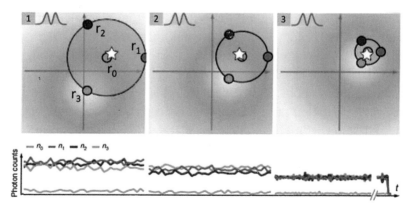

FIGURE 6.5 Principle of MINFLUX. The doughnut-shaped illumination pattern (green) is sequentially placed at the four positions $r_{0,1,2,3}$ indicated by blue, violet, red, and yellow dots, respectively (panel 1). The number of emitted photons $n_{0,1,2,3}$ counted for each doughnut position are used to extract the molecular location. To improve the localization precision, the scan range of the doughnuts can be reduced iteratively until the desired localization precision is attained (panels 2–3). (Adapted with permission from [64]. Copyright 2020 Springer Nature.)

all collected photons, MINFLUX finds the location of the doughnut spot that results in the minimal number of emitted photons. In this way, every detected photon tells you that the doughnut is not yet located exactly at the fluorophore's position. Reymond *et al.* described accurately the paradigm shift as going from "where is the fluorophore in the FOV?", for homogeneous illumination, to "is it located in the valley?" for MINFLUX [65].

Even though MINFLUX can obtain the same localization precision as standard localization methods, for 22 times fewer photons or for bringing it down to 3–5 nm with the same number of photons, it has two major disadvantages. First of all, MINFLUX has a limited field of view dictated by the choice of L, and secondly, it has a low throughput since every fluorophore is localized one-by-one with the triangulation approach. Multiple methods recently emerged that overcome these drawbacks and extend the MINFLUX principle for faster [66] and multicolor 3D acquisition [64]. MINFLUX and variants are further described in Chapter 7.

6.3.2 Modulation Enhanced Localization Microscopy

The principle of modulation enhanced localization microscopy (MELM), of which MINFLUX was the first example, is to use structured illumination as a ruler for probing the fluorophore's position [65]. The requirements of MELM are a well-defined illumination pattern, the ability to shift the pattern with high accuracy, and detectors that are sensitive and precise for single/low photon counts.

In a short period of time, multiple research groups came up with similar MELM ideas for the "parallelization" of MINFLUX to the entire FOV by using shifting and rotating sinusoidal illumination patterns [67–71]. In this way, the single-photon detector of MINFLUX is replaced by a camera and the doughnut illumination by a FOV-wide sinusoidal pattern.

Before going into the differences between these approaches, let us first introduce the principle with a 1D example (Figure 6.6). The sample is illuminated with a sinusoidal pattern. The fluorophore creates a diffraction-limited

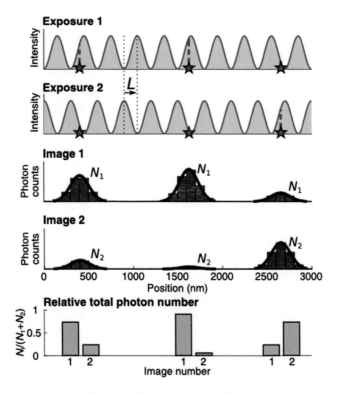

FIGURE 6.6   Principle of MELM. Under sinusoidal illumination (green), the average number of photons, N_1, in the diffraction-limited spot of each point-source (purple stars) depends on the position of the source relative to the illumination. If the structured illumination is phase-shifted a distance L, the average number of photons, N_2, changes. The changing relative photon numbers provide extra information for localizing the fluorophores. (Adapted with permission under a Creative Commons CC-BY 4.0 license from [66]. Copyright The Authors.)

spot on the camera, where the intensity of the spot depends on its position with respect to the sinusoidal pattern. By translating the pattern to another position, each spot remains at the same position, but with a different intensity. By fitting a sine function to the measured relative intensities of a spot corresponding to each position of the pattern, the position of the fluorophore is obtained relative to the illumination pattern. The problem that remains is that the fluorophore is localized *relative* to the illumination pattern, so since the pattern repeats itself over the FOV, the fluorophore could be located at every multiple of the pattern's periodicity (referred to as the modulo-problem). By rotating the sinusoidal pattern 90°, the y-position of the fluorophore is obtained in the same manner.

The first parallelization of MINFLUX was implemented as SIMPLE by Reymond et al. [70]. Here, the authors use phase-shifted sinusoidal patterns with three phase shifts and two orthogonal orientations, resulting in six intensity values per spot. The x-y localization with respect to the sinusoidal pattern is purely based on these six intensity values, where the modulo-problem is solved by using the rough PSF position.

In contrast to only considering the intensities, ROSE [68] and SIMFLUX [67] are improvements to SIMPLE that utilize the relative intensities and the position of the PSF combined to localize the fluorophore, which results in more robust estimation. Both methods report a near two-fold improvement in localization precision. The structured illumination is calibrated on the sample itself, which has the downside that the calibration uses photons that otherwise could have aided in improving the localization precision.

The C-MELM approach [66] overcomes this limitation by calibrating the structured illumination before it is used. This simplifies the analysis to just a single MLE fit to the joint collection of six images, resulting from 2 orientations and 3 phase shifts of the sinusoidal pattern. In this way, the single fit combines the information from the PSF shapes and intensities into one estimate of the lateral position of the fluorophore, and an improvement factor in localization precision of 2.1 is achieved.

Interestingly, the same logic from improving the localization precision using structured illumination holds in the axial direction, as shown with the ModLoc principle of Jouchet et al. [69]. The authors use the interference between two tilted plane waves to create interference fringes along the axial direction. Analogous to the lateral localization discussed above, the axial position of the fluorophore is obtained by measuring the intensity of spots for multiple phase-shifted interference fringes. The modulo-problem is solved by getting a rough axial estimate using astigmatic imaging and the 2D position is obtained by centroid fitting of the PSF.

6.4 OUTLOOK

The development of the above-mentioned techniques in the last decades has made it possible to study the position and orientation of single molecules within materials and biological tissues with high precision and accuracy. The future challenges are to improve these techniques to obtain a higher spatial resolution, a higher temporal resolution and reduced photodamage to the sample during imaging. We will highlight two promising recent developments that will enhance the field of SMLM: light-sheet microscopy and deep learning.

In light-sheet microscopy, the sample is illuminated with a thin sheet of light oriented perpendicular to the imaging objective. This approach of optical sectioning improves the signal-to-noise since it lowers the background fluorescence. Furthermore, the sample is illuminated with a lower intensity of light, thereby reducing the photodamage. Recently, light-sheet microscopy is combined with various localization microscopy approaches for imaging thick biological samples [72, 73]. A second promising development is the use of deep learning, which are artificial neural networks that can extract complex features from data after having been trained on a set of data. Deep learning has shown to enhance multiple aspects of localization microscopy pipelines. It can be used to extract positional and orientational information about the fluorophore from experimentally obtained images of isolated spots [74], replacing the need for PSF fitting approaches. Furthermore, it can be employed to localize multiple emitters from raw images with a high density of blinking fluorophores [75], and accelerate the acquisition by reconstructing the super-resolution images from fewer localizations [76].

These promising developments will bring the field of SMLM forward, allowing one to study the structure and function of materials and living systems with molecular detail.

REFERENCES

1. M. Born, E. Wolf, A. B. Bhatia, P. C. Clemmow, D. Gabor, A. R. Stokes, A. M. Taylor, P. A. Wayman, and W. L. Wilcock, *Principles of Optics: Electromagnetic Theory of Propagation, Interference and Diffraction of Light*, 7th ed. (Cambridge University Press, 1999).

2. E. Betzig, G. H. Patterson, R. Sougrat, O. W. Lindwasser, S. Olenych, J. S. Bonifacino, M. W. Davidson, J. Lippincott-Schwartz, and H. F. Hess, "Imaging intracellular fluorescent proteins at nanometer resolution," *Science* **313**, 1642–1645 (2006).

3. C. S. Smith, N. Joseph, B. Rieger, and K. A. Lidke, "Fast, single-molecule localization that achieves theoretically minimum uncertainty," *Nat. Methods* **7**, 373–375 (2010).

4. K. I. Mortensen, L. S. Churchman, J. A. Spudich, and H. Flyvbjerg, "Optimized localization analysis for single-molecule tracking and super-resolution microscopy," *Nat. Methods* **7**, 377–381 (2010).

5. J. D. Jackson, *Classical Electrodynamics* (John Wiley & Sons, Inc., 1999).

6. S. Stallinga and B. Rieger, "Accuracy of the Gaussian point spread function model in 2D localization microscopy," *Opt. Express* **18**, 24461 (2010).

7. J. Enderlein, "Theoretical study of detection of a dipole emitter through an objective with high numerical aperture," *Opt. Lett.* **25**, 634 (2000).

8. J. Enderlein, E. Toprak, and P. R. Selvin, "Polarization effect on position accuracy of fluorophore localization," *Opt. Express* **14**, 8111 (2006).

9. J. Enderlein and M. Böhmer, "Influence of interface–dipole interactions on the efficiency of fluorescence light collection near surfaces," *Opt. Lett.* **28**, 941 (2003).

10. L. S. Churchman, H. Flyvbjerg, and J. A. Spudich, "A non-Gaussian distribution quantifies distances measured with fluorescence localization techniques," *Biophys. J.* **90**, 668–671 (2006).

11. J. Engelhardt, J. Keller, P. Hoyer, M. Reuss, T. Staudt, and S. W. Hell, "Molecular orientation affects localization accuracy in superresolution far-field fluorescence microscopy," *Nano Lett.* **11**, 209–213 (2011).

12. K. I. Mortensen, J. Sung, H. Flyvbjerg, and J. A. Spudich, "Optimized measurements of separations and angles between intra-molecular fluorescent markers," *Nat. Commun.* **6**, 8621 (2015).

13. H. Deschout, F. Cella Zanacchi, M. Mlodzianoski, A. Diaspro, J. Bewersdorf, S. T. Hess, and K. Braeckmans, "Precisely and accurately localizing single emitters in fluorescence microscopy," *Nat. Methods* **11**, 253–266 (2014).

14. H. Cramér, "A contribution to the theory of statistical estimation," *Scand. Actuar. J.* **1946**, 85–94 (1946).

15. R. J. Ober, S. Ram, and E. S. Ward, "Localization accuracy in single-molecule microscopy," *Biophys. J.* **86**, 1185–1200 (2004).

16. A. Small and S. Stahlheber, "Fluorophore localization algorithms for super-resolution microscopy," *Nat. Methods* **11**, 267–279 (2014).

17. K. I. Mortensen and H. Flyvbjerg, "'Calibration-on-the-spot': how to calibrate an EMCCD camera from its images," *Sci. Rep.* **6**, 28680 (2016).

18. A. V. Abraham, S. Ram, J. Chao, E. S. Ward, and R. J. Ober, "Quantitative study of single molecule location estimation techniques," *Opt. Express* **17**, 23352 (2009).

19. M. Wang, J. M. Marr, M. Davanco, J. W. Gilman, and J. A. Liddle, "Nanoscale deformation in polymers revealed by single-molecule super-resolution localization–orientation microscopy," *Mater. Horiz.* **6**, 817–825 (2019).

20. L. Von Diezmann, Y. Shechtman, and W. E. Moerner, "Three-dimensional localization of single molecules for super-resolution imaging and single-particle tracking," *Chem. Rev.* **117**, 7244–7275 (2017).

21. C. N. Hulleman, R. Ø. Thorsen, E. Kim, C. Dekker, S. Stallinga, and B. Rieger, "Simultaneous orientation and 3D localization microscopy with a Vortex point spread function," *Nat. Commun.* **12**, 5934 (2021).

22. M. F. Juette, T. J. Gould, M. D. Lessard, M. J. Mlodzianoski, B. S. Nagpure, B. T. Bennett, S. T. Hess, and J. Bewersdorf, "Three-dimensional sub-100 nm resolution fluorescence microscopy of thick samples," *Nat. Methods* **5**, 527–529 (2008).

23. E. Toprak, H. Balci, B. H. Blehm, and P. R. Selvin, "Three-dimensional particle tracking via bifocal imaging," *Nano Lett.* **7**, 2043–2045 (2007).

24. P. Prabhat, S. Ram, E. S. Ward, and R. J. Ober, "Simultaneous imaging of different focal planes in fluorescence microscopy for the study of cellular dynamics in three dimensions," *IEEE Trans. Nanobioscience* **3**, 237–242 (2004).

25. S. Ram, D. Kim, R. J. Ober, and E. S. Ward, "3D single molecule tracking with multifocal plane microscopy reveals rapid intercellular transferrin transport at epithelial cell barriers," *Biophys. J.* **103**, 1594–1603 (2012).

26. L. Holtzer, T. Meckel, and T. Schmidt, "Nanometric three-dimensional tracking of individual quantum dots in cells," *Appl. Phys. Lett.* **90**, 053902 (2007).

27. B. Huang, W. Wang, M. Bates, and X. Zhuang, "Three-dimensional super-resolution imaging by stochastic optical reconstruction microscopy," *Science* **319**, 810–813 (2008).

28. H. P. Kao and A. S. Verkman, "Tracking of single fluorescent particles in three dimensions: use of cylindrical optics to encode particle position," *Biophys. J.* **67**, 1291–1300 (1994).

29. M. P. Backlund, M. D. Lew, A. S. Backer, S. J. Sahl, G. Grover, A. Agrawal, R. Piestun, and W. E. Moerner, "Simultaneous, accurate measurement of the 3D position and orientation of single molecules," *Proc. Natl. Acad. Sci.* **109**, 19087–19092 (2012).

30. S. R. P. Pavani and R. Piestun, "High-efficiency rotating point spread functions," *Opt. Express* **16**, 3484 (2008).

31. S. R. P. Pavani and R. Piestun, "Three dimensional tracking of fluorescent microparticles using a photon-limited double-helix response system," *Opt. Express* **16**, 22048 (2008).

32. R. Piestun, Y. Y. Schechner, and J. Shamir, "Propagation-invariant wave fields with finite energy," *J. Opt. Soc. Am. A* **17**, 294 (2000).

33. S. R. P. Pavani, M. A. Thompson, J. S. Biteen, S. J. Lord, N. Liu, R. J. Twieg, R. Piestun, and W. E. Moerner, "Three-dimensional, single-molecule fluorescence imaging beyond the diffraction limit by using a double-helix point spread function," *Proc. Natl. Acad. Sci.* **106**, 2995–2999 (2009).

34. M. D. Lew, S. F. Lee, M. Badieirostami, and W. E. Moerner, "Corkscrew point spread function for far-field three-dimensional nanoscale localization of pointlike objects," *Opt. Lett.* **36**, 202 (2011).

35. Y. Shechtman, S. J. Sahl, A. S. Backer, and W. E. Moerner, "Optimal point spread function design for 3D imaging," *Phys. Rev. Lett.* **113**, 133902 (2014).

36. Y. Shechtman, L. E. Weiss, A. S. Backer, S. J. Sahl, and W. E. Moerner, "Precise three-dimensional scan-free multiple-particle tracking over large axial ranges with tetrapod point spread functions," *Nano Lett.* **15**, 4194–4199 (2015).

37. A. P. Bartko and R. M. Dickson, "Imaging three-dimensional single molecule orientations," *J. Phys. Chem. B* **103**, 11237–11241 (1999).

38. M. Böhmer and J. Enderlein, "Orientation imaging of single molecules by wide-field epifluorescence microscopy," *J. Opt. Soc. Am. B* **20**, 554 (2003).

39. F. Aguet, S. Geissbühler, I. Märki, T. Lasser, and M. Unser, "Super-resolution orientation estimation and localization of fluorescent dipoles using 3-D steerable filters," *Opt. Express* **17**, 6829 (2009).

40. D. Patra, I. Gregor, and J. Enderlein, "Image analysis of defocused single-molecule images for three-dimensional molecule orientation studies," *J. Phys. Chem. A* **108**, 6836–6841 (2004).

41. M. Speidel, A. Jonáš, and E.-L. Florin, "Three-dimensional tracking of fluorescent nanoparticles with subnanometer precision by use of off-focus imaging," *Opt. Lett.* **28**, 69 (2003).

42. E. Toprak, J. Enderlein, S. Syed, S. A. McKinney, R. G. Petschek, T. Ha, Y. E. Goldman, and P. R. Selvin, "Defocused orientation and position imaging (DOPI) of myosin V," *Proc. Natl. Acad. Sci.* **103**, 6495–6499 (2006).

43. M. A. Lieb, J. M. Zavislan, and L. Novotny, "Single-molecule orientations determined by direct emission pattern imaging," *J. Opt. Soc. Am. B* **21**, 1210 (2004).

44. M. R. Foreman, C. M. Romero, and P. Török, "Determination of the three-dimensional orientation of single molecules," *Opt. Lett.* **33**, 1020 (2008).

45. Z. Sikorski and L. M. Davis, "Engineering the collected field for single-molecule orientation determination," *Opt. Express* **16**, 3660 (2008).

46. M. P. Backlund, M. D. Lew, A. S. Backer, S. J. Sahl, and W. E. Moerner, "The role of molecular dipole orientation in single-molecule fluorescence microscopy and implications for super-resolution imaging," *ChemPhysChem* **15**, 587–599 (2014).

47. A. S. Backer, M. Y. Lee, and W. E. Moerner, "Enhanced DNA imaging using super-resolution microscopy and simultaneous single-molecule orientation measurements," *Optica* **3**, 659 (2016).

48. T. J. Gould, M. S. Gunewardene, M. V. Gudheti, V. V. Verkhusha, S.-R. Yin, J. A. Gosse, and S. T. Hess, "Nanoscale imaging of molecular positions and anisotropies," *Nat. Methods* **5**, 1027–1030 (2008).

49. J. T. Fourkas, "Rapid determination of the three-dimensional orientation of single molecules," *Opt. Lett.* **26**, 211 (2001).

50. C. V. Rimoli, C. A. Valades-Cruz, V. Curcio, M. Mavrakis, and S. Brasselet, "4polar-STORM polarized super-resolution imaging of actin filament organization in cells," *Nat. Commun.* **13**, 1–13 (2022).

51. S. Stallinga and B. Rieger, "Position and orientation estimation of fixed dipole emitters using an effective Hermite point spread function model," *Opt. Express* **20**, 5896 (2012).

52. J. N. Forkey, M. E. Quinlan, M. Alexander Shaw, J. E. T. Corrie, and Y. E. Goldman, "Three-dimensional structural dynamics of myosin V by single-molecule fluorescence polarization," *Nature* **422**, 399–404 (2003).

53. J. N. Forkey, M. E. Quinlan, and Y. E. Goldman, "Measurement of single macromolecule orientation by total internal reflection fluorescence polarization microscopy," *Biophys. J.* **89**, 1261–1271 (2005).

54. A. S. Backer, M. P. Backlund, M. D. Lew, and W. E. Moerner, "Single-molecule orientation measurements with a quadrated pupil," *Opt. Lett.* **38**, 1521 (2013).

55. A. S. Backer, M. P. Backlund, L. Von Diezmann, S. J. Sahl, and W. E. Moerner, "A bisected pupil for studying single-molecule orientational dynamics and its application to three-dimensional super-resolution microscopy," *Appl Phys Lett* **104**, 193701 (2014).

56. O. Zhang, J. Lu, T. Ding, and M. D. Lew, "Imaging the three-dimensional orientation and rotational mobility of fluorescent emitters using the Tri-spot point spread function," *Appl. Phys. Lett.* **113**, 031103 (2018).

57. T. Ding and M. D. Lew, "Single-molecule localization microscopy of 3D orientation and anisotropic wobble using a polarized Vortex point spread function," *J. Phys. Chem. B* **125**, 12718–12729 (2021).

58. H. Heydarian, F. Schueder, M. T. Strauss, B. Van Werkhoven, M. Fazel, K. A. Lidke, R. Jungmann, S. Stallinga, and B. Rieger, "Template-free 2D particle fusion in localization microscopy," *Nat. Methods* **15**, 781–784 (2018).

59. T.A.P.M. Huijben, H. Heydarian, A. Auer, F. Schueder, R. Jungmann, S. Stallinga, and B. Rieger, "Detecting structural heterogeneity in single-molecule localization microscopy data," *Nat.Commun.* 12, 1–8 (2021). https://doi.org/10.1038/s41467-021-24106-8

60. J. B. Grimm, B. P. English, J. Chen, J. P. Slaughter, Z. Zhang, A. Revyakin, R. Patel, J. J. Macklin, D. Normanno, R. H. Singer, T. Lionnet, and L. D. Lavis, "A general method to improve fluorophores for live-cell and single-molecule microscopy," *Nat. Methods* **12**, 244–250 (2015).

61. S. Weisenburger, D. Boening, B. Schomburg, K. Giller, S. Becker, C. Griesinger, and V. Sandoghdar, "Cryogenic optical localization provides 3D protein structure data with Angstrom resolution," *Nat. Methods* **14**, 141–144 (2017).

62. A. Puchkova, C. Vietz, E. Pibiri, B. Wünsch, M. Sanz Paz, G. P. Acuna, and P. Tinnefeld, "DNA Origami nanoantennas with over 5000-fold fluorescence enhancement and single-molecule detection at 25 µm," *Nano Lett.* **15**, 8354–8359 (2015).

63. F. Balzarotti, Y. Eilers, K. C. Gwosch, A. H. Gynnå, V. Westphal, F. D. Stefani, J. Elf, and S. W. Hell, "Nanometer resolution imaging and tracking of fluorescent molecules with minimal photon fluxes," *Science* **355**, 606–612 (2017).

64. K. C. Gwosch, J. K. Pape, F. Balzarotti, P. Hoess, J. Ellenberg, J. Ries, and S. W. Hell, "MINFLUX nanoscopy delivers 3D multicolor nanometer resolution in cells," *Nat. Methods* **17**, 217–224 (2020).

65. L. Reymond, T. Huser, V. Ruprecht, and S. Wieser, "Modulation-enhanced localization microscopy," *J. Phys. Photonics* **2**, 041001 (2020).
66. M. Schmidt, A. C. Hundahl, H. Flyvbjerg, R. Marie, and K. I. Mortensen, "Camera-based localization microscopy optimized with calibrated structured illumination," *Commun. Phys.* **4**, 41 (2021).
67. J. Cnossen, T. Hinsdale, R. Ø. Thorsen, M. Siemons, F. Schueder, R. Jungmann, C. S. Smith, B. Rieger, and S. Stallinga, "Localization microscopy at doubled precision with patterned illumination," *Nat. Methods* **17**, 59–63 (2020).
68. L. Gu, Y. Li, S. Zhang, Y. Xue, W. Li, D. Li, T. Xu, and W. Ji, "Molecular resolution imaging by repetitive optical selective exposure," *Nat. Methods* **16**, 1114–1118 (2019).
69. P. Jouchet, C. Cabriel, N. Bourg, M. Bardou, C. Poüs, E. Fort, and S. Lévêque-Fort, "Nanometric axial localization of single fluorescent molecules with modulated excitation," *Nat. Photonics* **15**, 297–304 (2021).
70. L. Reymond, J. Ziegler, C. Knapp, F.-C. Wang, T. Huser, V. Ruprecht, and S. Wieser, "SIMPLE: structured illumination based point localization estimator with enhanced precision," *Opt. Express* **27**, 24578 (2019).
71. R. Schmidt, T. Weihs, C. A. Wurm, I. Jansen, J. Rehman, S. J. Sahl, and S. W. Hell, "MINFLUX nanometer-scale 3D imaging and microsecond-range tracking on a common fluorescence microscope," *Nat. Commun.* **12**, 1478 (2021).
72. M. J. Mlodzianoski, P. J. Cheng-Hathaway, S. M. Bemiller, T. J. McCray, S. Liu, D. A. Miller, B. T. Lamb, G. E. Landreth, and F. Huang, "Active PSF shaping and adaptive optics enable volumetric localization microscopy through brain sections," *Nat. Methods* **15**, 583–586 (2018).
73. F. Wäldchen, J. Schlegel, R. Götz, M. Luciano, M. Schnermann, S. Doose, and M. Sauer, "Whole-cell imaging of plasma membrane receptors by 3D lattice light-sheet dSTORM," *Nat. Commun.* **11**, 887 (2020).
74. P. Zhang, S. Liu, A. Chaurasia, D. Ma, M. J. Mlodzianoski, E. Culurciello, and F. Huang, "Analyzing complex single-molecule emission patterns with deep learning," *Nat. Methods* **15**, 913–916 (2018).
75. E. Nehme, D. Freedman, R. Gordon, B. Ferdman, L. E. Weiss, O. Alalouf, T. Naor, R. Orange, T. Michaeli, and Y. Shechtman, "DeepSTORM3D: dense 3D localization microscopy and PSF design by deep learning," *Nat. Methods* **17**, 734–740 (2020).
76. W. Ouyang, A. Aristov, M. Lelek, X. Hao, and C. Zimmer, "Deep learning massively accelerates super-resolution localization microscopy," *Nat. Biotechnol.* **36**, 460–468 (2018).

CHAPTER 7

Single Molecule Localization and Nanoscopy Through Sequential Structured Illumination

Piotr Zdańkowski

Consejo Nacional de Investigaciones Científicas y Técnicas (CONICET), Buenos Aires, Argentina

Warsaw University of Technology, Institute of Micromechanics and Photonics, Warsaw, Poland

Lucía F. Lopez

Consejo Nacional de Investigaciones Científicas y Técnicas (CONICET), Buenos Aires, Argentina

Florencia Edorna

Consejo Nacional de Investigaciones Científicas y Técnicas (CONICET), Buenos Aires, Argentina

Universidad de Buenos Aires, Ciudad Autónoma de Buenos Aires, Argentina

Guillermo P. Acuna

University of Fribourg, Fribourg, Switzerland

DOI: 10.1201/9781003220688-7

Fernando D. Stefani

Consejo Nacional de Investigaciones Científicas y Técnicas (CONICET), Buenos Aires, Argentina

Universidad de Buenos Aires, Ciudad Autónoma de Buenos Aires, Argentina

The localization precision of single fluorophores, as well as the resolution achieved with single-molecule localization microscopy (SMLM) or coordinate-targeted super-resolution methods, is ultimately limited by the photostability of fluorophores [1–3]. As described in Chapter 2, under ambient conditions, the typical lateral resolution lies in the range of 10–40 nm, whereas the axial resolution is usually 3–5 times worse. Hence, recently, efforts have been devoted to developing methods enabling fluorescence imaging at the 1–10 nm scale, which would provide true molecular resolution by surpassing the typical size of structural proteins and even reaching the size of the fluorophores. The photostability limitation can be bypassed by two approaches: through obtaining more fluorescence photons from a given position in the sample, or by extracting more information from the limited photon budget. Both avenues have been explored to achieve super-resolution imaging with sub-10 nm resolution. DNA-PAINT (DNA-Point Accumulation for Imaging in Nanoscale Topography) is a SMLM method that uses transient hybridization of fluorescently labeled single-stranded DNA sequences (imager strands) to the target molecules/positions of the sample, which are functionalized and mediated with complementary DNA sequences (docking strands). This enables constant inspection of the target molecule/position with multiple fluorophores with nearly unlimited fluorescent photons [4, 5]. Dynamic binding-unbinding of the imager and docking strands generates the necessary blinking for SMLM. The advantage of DNA-PAINT lies in the fact that the virtually unlimited localization events can be measured for each labeled position of the sample, which can lead to sub 10 nm resolution [6–8]. A second approach aims to extract more information about the molecular position from the limited photon budget using a sequence of spatially modulated excitation beams. The latter approach for the ultra-precise localization of single fluorophores was pioneered by the so-called MINFLUX concept [9–11] which uses beams comprising a minimum of light.

MINFLUX stands for MINimal emission FLUXes, since this technique aims to use minimal photon counts for reaching a given single molecule localization precision. The localization precision of MINFLUX can be increased at the expense of reducing the field of view, and can reach the 1 nm limit routinely, which constitutes a 10-fold improvement when compared with other SMLM methods based on wide field camera detection. Since its original publication, MINFLUX has proved to be able to attain single-molecule resolution in DNA origami structures, fixed and living biological cells both in 2- and 3-dimensions [9–15], in principle, without constraints from wavelength, numerical aperture (NA), or molecular orientation.

In this chapter, we will first describe the working principle of MINFLUX, and then provide a conceptual framework useful to describe and benchmark MINFLUX with other methods that work under the same principle. In fact, a whole family of methods for single-molecule localization using sequential structured illumination has been developed over the years. The first one of them, the so-called orbital tracking (OT) method, was proposed theoretically more than 20 years ago [16], and was later applied for the tracking of single molecules in 2D and 3D. Until recently, OT, MINSTED, [17] and other methods were considered to be unrelated, while in fact they can all be understood with a common framework to MINFLUX. Furthermore, the common framework serves to explore and evaluate new concepts, as it was done with the recently introduced RASTMIN (RASTer scanning a MINimum) method, which provides localization precision equivalent to MINFLUX using a conventional scanning (confocal) microscope with only two modifications.

7.1 MINFLUX

In MINFLUX, the position of a single fluorescent molecule is obtained from a sequence of exposures to a focused laser beam that features an intensity minimum. For 2D localization, four exposures to a toroidal (donut) focus are used. The excitation beam has two main roles: it serves both as a probe to excite the fluorescence, as well as a reference coordinate for the localization. The position and intensity profile of the beam define how many photons should be detected, on average, from a single molecule at a given position (e.g. the closer the fluorophore is to the intensity minimum, the fewer photons are detected). Reciprocally, the molecular position may be estimated from the set of detected photons for each position of the beam. It turns out that performing this kind of estimation using beams comprising a minimum (ideally a zero) of intensity has two important advantages: it is very photon efficient and it enables indefinite zooming in to achieve higher localization precision at the expense of reducing the field of view. That is where the comparative advantage of MINFLUX lies.

The concept is easily understood in 1D. Let us first assume having a beam centered at a position x_b with a one-dimensional (1D) intensity profile $I(x - x_b)$, as shown in Figure 7.1b. In general, exciting a molecule located at x_m with such a profile will result in a fluorescence readout, expressed in number of photons n given by:

$$n(x_m) = A * I(x_m - x_b), \tag{7.1}$$

where A is a factor proportional to the brightness and orientation of the molecule, as well as to the detection efficiency. We now consider $I(x)$ comprising a minimum (ideally a zero) of intensity. Near that minimum, $I(x)$ may be well-approximated by a parabola. In principle, for the 1D localization problem, only two measurements with $I(x)$ displaced by distance L are sufficient. If L is sufficiently small enough to justify a parabolic approximation of $I(x)$, the two detected fluorescence signals will be:

$$n_0(x_m) = A\, I\left(x_m - \frac{L}{2}\right) = A'\left(x_m - \frac{L}{2}\right)^2$$

$$n_1(x_m) = A\, I\left(x_m + \frac{L}{2}\right) = A'\left(x_m + \frac{L}{2}\right)^2 \tag{7.2}$$

where A' includes a constant proportional to the total laser intensity. The acquired photon counts n_0 and n_1 for the two probing positions follow Poissonian statistics with means corresponding to the local excitation intensity. The probability of acquiring n_0 in the first measurement, given that a total of $N = n_0 + n_1$ photons are detected, can be expressed with a binomial probability distribution $P(p_0, N)$, where p_0 can be expressed as:

$$p_0(x) = \frac{n_0(x)}{n_0(x) + n_1(x)} = \frac{\left(1 + \dfrac{2x}{L}\right)^2}{2\left(1 + \dfrac{4x^2}{L^2}\right)}. \tag{7.3}$$

Similarly, p_1 can be expressed as:

$$p_1(x) = \frac{n_1(x)}{n_0(x) + n_1(x)} = \frac{\left(1 - \dfrac{2x}{L}\right)^2}{2\left(1 + \dfrac{4x^2}{L^2}\right)}. \tag{7.4}$$

Knowing the success probability, a likelihood function $\mathcal{L}(p\,|\,\bar{n})$ can be expressed as:

$$\mathcal{L}(x\,|\,n) = \frac{N!}{n_0!\,n_1!} \frac{\left(1 + \dfrac{2x}{L}\right)^{2n_0}\left(1 - \dfrac{2x}{L}\right)^{2n_1}}{4\left(1 + \dfrac{4x^2}{L^2}\right)^{2N}}, \tag{7.5}$$

considering $N = n_0 + n_1$. Finally, maximum likelihood estimation (MLE) can be used to estimate the position x_m of the single molecule from the photon collection and the probabilities:

$$\bar{x}_m^{\text{MLE}}(n_0, N) = -\frac{L}{2} + \frac{L}{1 + \sqrt{\dfrac{n_1}{n_0}}}, \tag{7.6}$$

for which $p_0(\hat{x}_m) = \dfrac{n_0}{n_0 + n_1}$. Such a statistical description of the MINFLUX method allows the calculation of the Cramér-Rao bound (CRB) governing the best possible localization precision that is achieved with any unbiased estimator, which for the 1D case is: of standing wave is:

$$\sigma_{\text{CRB}}(x) = \frac{L}{4\sqrt{N}}\left(1 + \frac{4x^2}{L^2}\right). \tag{7.7}$$

In MINFLUX, the CRB reaches its minima for the center of the field of view (FOV):

$$\sigma_{\text{CRB}}(0) = \frac{L}{4\sqrt{N}}, \tag{7.8}$$

This 1D result based on the parabolic approximation already shows the key advantages of MINFLUX: the localization precision is independent of the wavelength and can be improved at the expense of reducing the "field of view" L. In other words, under ideal conditions, one could zoom in indefinitely. As an example, with $L = 150$ nm a localization precision better than 4 nm is achieved with just 100 photons. Reducing L to 100 nm increases the precision to 2.5 nm. In practice, background photons impose a limitation to reduce L indefinitely. Iterative zooming in using the previous localization information makes the localization procedure even more efficient in terms of the use of photons [10].

All the considerations above can be extended to 2D and any illumination intensity profile that displays a central zero. In the original concept, [9] a donut beam with an intensity zero in the center was used for 2D localization, Figure 7.1a. 2D MINFLUX with a donut illumination requires registration of photon counts from four positions, three at the vertices of an equilateral triangle and one at the center. The molecular position is estimated from the set of four photon counts n_0, n_1, n_2, n_3, the positions of the beam defining L (Figure 7.1c), and the spatial illumination profile $I(x, y)$ which is obtained from a high resolution, high Signal-to-Noise Ratio (SNR), image of the point spread function (PSF). Similarly to the 1D case, the CRB is linearly dependent of L and wavelength independent [9].

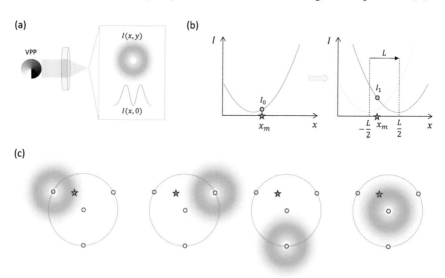

FIGURE 7.1 (a) Donut shaped excitation beam using a vortex phase plate. (b) Single molecule localization scheme for the 1D localization problem. (c) 2D localization, registration of photon counts from four positions.

MINFLUX has been extended to achieve 3D localization of single molecules [10, 12]. This requires an excitation beam with a three-dimensional intensity zero (so-called 3D donut), just as it is used for depletion in 3D STED microscopy [18–20]. 3D MINFLUX localization requires two additional measurements with the excitation beam in axially displaced positions. The measurement pipeline for 3D localization proposed by the 3D MINFLUX method [10] can be summarized as follows: first, the three-dimensional position (x, y, z) of the molecule is roughly determined using a Gaussian beam. Next, similarly to the 1D localization problem described above, the axial position (z) of the fluorophore is estimated from two measurements performed with the 3D donut, one below and the other above the estimated position of the fluorophore. The last step is an iterative process where the 3D position of the fluorophore is being estimated with increasing precision by performing 6 measurements with the 3D donut (4 for the lateral position and 2 for the axial position) at sequentially smaller separations. Each iteration step uses the previous estimation of the molecular position to re-center the excitation pattern. Such a localization approach with a 3D donut results in isotropic precision of ~1 nm with $N = 1000$ detected photons covering a volume of 400 nm diameter [10]. 3D MINFLUX, so far has helped to discern the cyto- and nucleoplasmic layers of Nup96 in single nuclear pore complexes (which are usually 50 nm apart in the axial direction) with ~2 nm precision and showed the multicolor imaging capabilities, [10, 12] and has enabled single fluorophore 3D tracking with <20 nm precision and 100 µs temporal resolution to study the diffusion of single lipids in lipid-bilayer model membranes [12]. More recently, MINFLUX also allowed for 3D tracking and direct observation of the stepping motion of the motor protein kinesin-1 in living cells with <2 nm and <1 ms spatiotemporal resolution [13, 15].

7.1.1 A Common Framework

MINFLUX in all its variants, along with several other reported methods for single-molecule localization can be understood within a common framework of Single Molecule Localization by Sequential Structured Illumination (SML-SSI) [21].

Figure 7.2 shows the general concept of SML-SSI with different types of excitation beams and scanning patterns. A spatially structured light with intensity $I(\mathbf{r})$ is sequentially shifted along K positions $\mathbf{r}_i$ ($1 \leq i \leq K$). Let us call the sequence of $I(\mathbf{r} - \mathbf{r}_i)$ the excitation beam pattern (EBP) and $\mathbf{r}_m$ the sought position of the emitter. Chosen positions $\mathbf{r}_i$ must not be

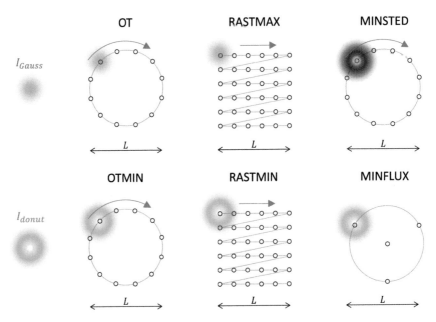

FIGURE 7.2   Different configurations of SML-SSI methods using focused beams with a central maximum (I_{Gauss}) or with a central minimum (I_{donut}).

co-linear so that localization ambiguities are avoided. Apart from that, the positions can be arbitrary in the plane. We assume a linear fluorescence excitation. The excited molecule will emit a number of fluorescence photons according to the local intensity of $I(r_m - r_i)$. The registered number of photon counts n_i are assumed to have a Poisson distribution. The position of the molecule is then estimated from the series of intensity measurements n_i, and using the knowledge of $I(r)$ and r_i.

Experimentally, performing a SML-SSI measurement usually consists of two steps: i) a highly detailed measurement of $I(r)$ with high Signal-to-Noise Ratio (SNR) using sufficiently bright emitters (ideally enabling photon counts in the range of $N > 10^6$) such as fluorescent nanoparticles, and ii) the single molecule measurements of n_i. Here, for simplicity, we will describe $I(r)$ analytically and consider n_i will be simulated from a Poisson distribution. We will consider SML-SSI methods using focused laser beams, which can be divided into two groups:

- beams with a central maximum of intensity (e.g., Gaussian beam),

- beams with a central minimum of intensity (e.g., donut beam).

We will describe the beam intensity profile with a central maximum using a Gaussian function:

$$I_{\text{Gauss}}\left(r\right) = A_0 e^{-4\ln\left(\frac{2r^2}{\text{FWHM}^2}\right)}. \tag{7.9}$$

The intensity profile with a central zero, the donut beam, will be described as:

$$I_{\text{donut}}\left(r\right) = A_0 4e\ln\left(\frac{2r^2}{\text{FWHM}^2}\right)e^{-4\ln\left(\frac{2r^2}{\text{FWHM}^2}\right)}. \tag{7.10}$$

Two types of EBPs will be considered: orbital sequences and raster scanning sequences. However, different combinations of illumination schemes with $I_{\text{Gauss}}(r)$ and $I_{\text{donut}}(r)$ can be used in SML-SSI. Figure 7.2 shows the different combinations of illumination and molecule excitation sequences. The total number of exposures K may also vary from just a few (e.g., 4 in MINFLUX [9, 11]) to an almost continuous scan in OT [16]. The recently reported MINSTED [17] method is essentially an extension of OT with a sub-diffraction effective $I_{\text{Gauss}}(r)$ obtained through STED, resulting in increased localization precision. OT could also be used with a central minimum intensity beam like $I_{\text{donut}}(r)$, although such an approach has not been reported (OTMIN). Lastly, SML-SSI can be implemented in a raster scanning microscope [21–24] with either $I_{\text{Gauss}}(r)$ (RASTMAX) or $I_{\text{donut}}(r)$ (RASTMIN).

The estimation of the molecular position from $[I(r), n_i, r_i]$ can be carried out using different approaches (e.g., using MLE in MINFLUX [9, 11], frequency analysis in Fourier space [25], intensity signal triangulation [26] or single particle localization using four foci [27, 28]). Here, we will use MLE as it is widely used; its performance is well-established, is consistent, and reaches asymptotically the Cramér-Rao bound (CRB) [29].

If we call $N = n_0 + \ldots + n_{K-1}$ the total number of detected photons, the probability of obtaining a certain combination of n_i, $\boldsymbol{n} = [n_1, n_2, \ldots, n_K]$, follows a multinomial probability distribution:

$$P(\boldsymbol{n}\,|\,N) = \frac{N!}{n_1!\cdots n_K!}\prod_{i=1}^{K} p_i\left(r_m\right)^{n_i}, \tag{7.11}$$

where the dependence on molecular position is included in the multinomial parameters $p_i(r_m)$:

$$p_i\left(r_m\right) = \frac{n_i\left(r_m - r_i\right)}{\sum_{j=1}^{K} n_j\left(r_m - r_j\right)}. \tag{7.12}$$

The likelihood function $\mathcal{L}$ for the position of the molecule can be described as follows:

$$\mathcal{L}(r_m \mid n) = \frac{N!}{\prod_{i=1}^{K} n_i!} \prod_{i=1}^{K} p_i\left(r_m\right)^{n_i}. \tag{7.13}$$

For the MLE calculation, the log-likelihood function is usually more practical, since the aim is to find $\hat{r}_m$ for which $\mathcal{L}(r_m \mid n) = \max$:

$$l(r_m \mid n) = \ln\left(\mathcal{L}(r_m \mid n)\right) = \sum_{i=1}^{K} n_i \ln\left[p_i(r_m \mid n)\right]. \tag{7.14}$$

The maximum likelihood estimator for the molecule position r_m can be defined as:

$$\hat{r}_m^{MLE} = argmax\, l\,(r_m \mid n). \tag{7.15}$$

In real experimental conditions, background will affect the readout, hence $p_i(r_m)$ will become [21]:

$$p_i\left(r_m\right) = \frac{SBR\left(r_m\right)}{SBR\left(r_m\right)+1} \frac{I\left(r_m - r_i\right)}{\sum_{j=1}^{K} I\left(r_m - r_i\right)} + \frac{1}{SBR\left(r_i\right)+1} \frac{1}{K}, \tag{7.16}$$

where SBR is the Signal-to-Background Ratio, defined as the ratio between the total photon counts from the molecule and the total background counts for all the exposures. SBR(r_m) depends on the molecular position because the total photon counts of the molecule depends on position. The total background counts are considered to be independent of the emitter position. We assume that the background counts have a Poisson distribution and are equal for all exposures. Other sources of background (non-Poissonian) or detection noise could be included but are not considered in this model.

Having the above constraints in mind, the SBR(r_m) can be expressed as a function of the SBR at the center of the excitation illumination intensity pattern SBR(0):

$$\text{SBR}\left(r_m\right) = \text{SBR}\left(0\right) \frac{\sum_{j=1}^{K} I\left(r_m - r_j\right)}{\sum_{j=1}^{K} I\left(0 - r_j\right)}. \tag{7.17}$$

SML-SSI methods can deliver high precision position estimation for the molecules that are in the vicinity of the EBP. The further they are from its center, the lower the localization precision will be. Some methods may actually become inviable for molecules far from the EBP. This problem can be easily tackled with the aid of prior information about the approximate emitter position, so that the EBP can be placed in the close neighborhood of the target molecule. The log-likelihood function can now be written as:

$$l(r_m \mid n) = \sum_{i=1}^{K} n_i \ln\left[p_i(r_m \mid n)\right] + \ln\left[f\left(r_m\right)\right], \tag{7.18}$$

where $f(r_m)$ holds the prior information about the rough emitter position.

The CRB, a theoretical lower bound for the variance of the chosen estimator, is calculated from the Fisher information matrix (a theoretical maximum precision matrix for an unbiased position estimator) [30]. Considering the 2D case of single molecule localization, the Fisher information matrix is:

$$\begin{aligned}
\mathcal{F}\left(r_m\right) &= \mathcal{F}_{\text{SML-SSI}} + \mathcal{F}_{\text{prior}} \\
&= \sum_{i=1}^{K} n_i \sum_{i=1}^{K} \frac{1}{p_i}
\begin{bmatrix}
\left(\dfrac{\partial p_i}{\partial x}\right)^2 & \dfrac{\partial p_i}{\partial x}\dfrac{\partial p_i}{\partial y} \\[2ex]
\dfrac{\partial p_i}{\partial y}\dfrac{\partial p_i}{\partial x} & \left(\dfrac{\partial p_i}{\partial y}\right)^2
\end{bmatrix}
-
\begin{bmatrix}
\dfrac{\partial^2 \ln f}{\partial x^2} & \dfrac{\partial^2 \ln f}{\partial x \partial y} \\[2ex]
\dfrac{\partial^2 \ln f}{\partial y \partial x} & \dfrac{\partial^2 \ln f}{\partial y^2}
\end{bmatrix}.
\end{aligned} \tag{7.19}$$

Having the Fisher estimation matrix, a Bayesian CRB, [31] the lower bound for the covariance matrix of the estimated molecule position $\Sigma_{\text{cov}}(r_m)$ can be obtained using the Cramér-Rao inequality:

$$\Sigma_{\text{cov}}\left(r_m\right) \geq \Sigma_{\text{CRB}}\left(r_m\right) = \mathcal{F}\left(r_m\right)^{-1}. \tag{7.20}$$

From Eq. 7.20, $\Sigma_{\text{CRB}}(r_m)$ can be calculated for any position of the molecule r_m. For simplicity, we will compute σ_{CRB}, the arithmetic mean of the eigenvalues of $\mathcal{F}\left(r_m\right)^{-1}$:

$$\sigma_{\text{CRB}}\left(r_m\right) = \sqrt{\frac{1}{D}\,\text{tr}\left[\Sigma_{\text{CRB}}\left(r_m\right)\right]}, \tag{7.21}$$

where D is the number of dimensions of the analyzed localization case (2 for the 2D case). Having the prior knowledge of the coarse position of molecule of interest, $f(r_m)$ helps to reduce uncertainty in the final position estimation.

Using this framework, a fair comparison of the performance of different SML-SSI approaches can be done. Figure 7.3 shows a comparison of best performances for OT, OTMIN, MINFLUX, RASTMIN, and RASTMAX [21] for $N = 500$ and SBR = 5. L was adapted for each method. Here is where all methods based on intensity minima outstand, as they provide the zooming-in capacity by reducing L well beyond the diffraction limit. Even though smaller L could be used, here the performance of all methods using a minimum (OTMIN, MINFLUX and RASTMIN) is shown for $L = 100$ nm. By contrast, methods using a maximum of intensity have a range of L for optimum performance that is somewhat determined by diffraction limit. In theory, for a Gaussian beam, the localization precision of OT increases with L. However, localization using the tails of a focused beam is not practical in reality as the beams are not really Gaussian in that range, and present secondary minima and maxima. For this reason, the performance of orbital tracking is best when the diameter of the orbit is similar to the half-width of the beam ($L = 300$ nm). For RASTMAX, the scanned area should be large enough to include the image of the molecule ($L = 600$ nm) (i.e., the size of the PSF). Increasing the scanned area beyond that does not provide significant information as mainly background would be detected. A more detailed comparison between methods can be found in our recent work by Masullo et al. [21]

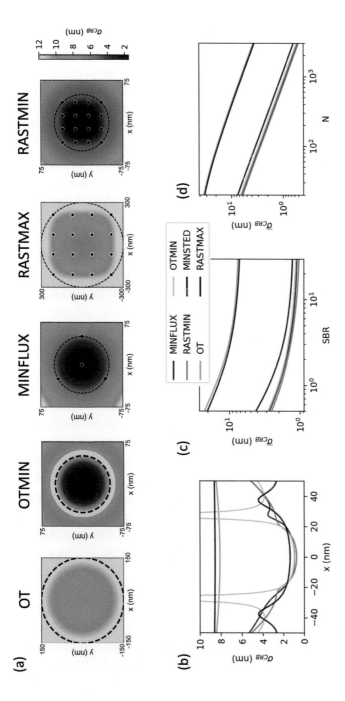

FIGURE 7.3 Comparison of different SML-SSI methods. (a) Localization precision maps of $\sigma_{CRB}(\mathbf{r}_m)$ for $N = 500$ SBR = 5, and optimum L for each method: orbital tracking (OT, $L = 300$ nm), orbital tracking with minimum (OTMIN, $L = 100$ nm), MINFLUX ($L = 100$ nm), RASTMIN ($L = 100$ nm), and RASTMAX ($L = 600$ nm). (b) Localization precision σ_{CRB} vs molecular position along x. (c) Average localization precision over a region of diameter of $0.75L$ ($\bar{\sigma}_{CRB}$) as a function of SBR. (d) $\bar{\sigma}_{CRB}$ as a function of N. $N = 500$ and SBR = 5 unless otherwise stated. (Adapted from Masullo et al. [21]. Copyright 2022 Elsevier.)

Figure 7.3a shows localization precision maps of $\sigma_{\text{CRB}}(r_m)$ for OT, OTMIN, MINFLUX, RASTMAX, and RASTMIN. All techniques show the best localization performance in the central region of the EBP, in an area defined by 75–80% of L, and methods using a light minimum are consistently better in terms of precision. OTMIN shows a region of low precision near the orbit due to the symmetry of the estimation problem in those positions. A more quantitative comparison $\sigma_{\text{CRB}}(r_m)$ is shown in the profiles of Figure 7.3b. Near the center of the EBP, methods using a light minimum are about one order of magnitude more precise than methods using a maximum. Beyond the size of the EBP, methods using a minimum become comparable, and then worse, than methods using a maximum. The only exception is MINSTED, which uses a sub-diffraction effective maximum (FWHM = 50 nm). As a function of the SBR and N, all methods present a similar dependency, as shown in Figure 7.3c and d, respectively. From the methods using diffraction limited excitation (all except MINSTED), the methods using a minimum are much more photon efficient. For example, reaching σ_{CRB} of 1 nm with OT or RASTMAX requires $N > 30,000$, while OTMIN, MINFLUX, or RASTMIN achieve this precision with $N \sim 1,000$ photons.

7.2 RASTMIN

From the comparison of different alternatives of SML-SSI techniques, RASTMIN [22] emerges as a simple solution to achieve molecular localization precision with minimal modifications to a common raster scanning fluorescence microscope such as a confocal or a two-photon excitation microscope. The schematic of its principle is shown in Figure 7.4a. RASTMIN uses an excitation intensity profile with a central zero, $I_{\text{donut}}(r)$, to scan the sample. As shown in the previous section, the expected localization precision for usual experimental conditions is around 1 nm in the center of the excitation pattern and does not exceed 5 nm over an area beyond the excitation pattern. Only two modifications are needed to convert a scanning microscope into a RASTMIN system. First, the addition of a vortex phase plate (or any other method of creating an intensity minimum, for example, spatial light modulator [20, 32] or segmented phase plate [33]) and polarization optics into the excitation beam path. The second modification is an active drift correction module, so that the sample remains stable at its lateral and axial position relative to the optical system. Ideally, such a drift correction system should be able to correct the sample position

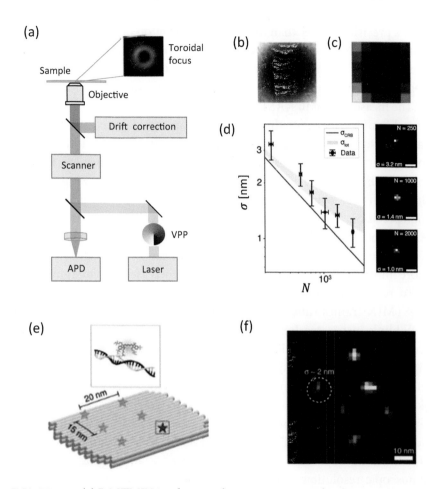

FIGURE 7.4 (a) RASTMIN implemented on a scanning confocal imaging system. (b) Image of the excitation beam obtained with high resolution and high SNR as reference measurement of $I(r)$. The white square marks a 100×100 nm^2 area. (c) 100×100 nm^2, 6×6 pixels image of the excitation beam and obtained with a single molecule to determine its position. (d) Experimental and theoretical localization error. The red curve shows the σ_{CRB}. The grey area denotes the expected precision considering the σ_{CRB} and the experimentally determined precision of the drift correction system. On the right, histograms of localizations obtained with different N. (e) and (f) show the design of the DNA Origami test sample and the reconstructed image with $\sigma_{CRB} < 2$ nm precision, respectively. (Adapted with permission under a Creative Commons CC-BY 4.0 license from Masullo et al. [22]. Copyright 2022 The Authors.)

with a precision higher than the desired localization precision. It should be noted that although RASTMIN could be performed with any type of photodetectors, it is convenient to use single photon counting detectors with negligible dark-counts and readout noise such as avalanche photodiodes.

The measurement in RASTMIN starts by acquisition of the high resolution, high SBR image of the excitation beam $I(r)$, Figure 7.4b. Then, the sample is scanned with the excitation beam over the region of interest using low power illumination, so that the individual molecules are initially located. The initial localization is followed by the raster scanning of a sub-diffraction area over the target molecule to determine its position with high precision. These sub-diffraction images are called RASTMIN frames. An example of a single frame is shown in Figure 7.4c. It turns out that using $K = 6 \times 6 = 36$ pixels is a good experimental choice. While larger K provides more information, the localization precision increases only marginally.

To reconstruct a super-resolved image using RASTMIN, a series of RASTMIN frames should be acquired while single molecules inside the FOV blink, just as it is done in MINFLUX or SMLM. An exemplary nanoscopic image reconstruction of a DNA origami structure can be seen in Figure 7.4e, f, where a rectangular arrangement of fluorophores with 3×2 sites of DNA origami is imaged, with an experimental localization precision of ~2 nm [22].

RASTMIN can successfully localize single molecules with 1 nm precision using low excitation intensity, just as MINFLUX can. When combined with fluorophore blinking, it can be used as an imaging device with nanoscopic resolution. What is unique about this method, when compared to other SML-SSI techniques providing similar localization precision, is that it can be relatively easily implemented in any laser-scanning microscope. It provides a way for many labs around the world that already count with a confocal or two-photon microscope to jump into the super-resolution field with true molecular resolution. Furthermore, RASTMIN has the potential to be parallelized with full-field structured illumination with, for example, an array of intensity minima [34] and extended to 3D single molecule localization with the addition of a top-hat phase mask [10] (similarly to 3D STED microscopy), 4Pi configuration [35], SIMPLER approach [36], or PSF engineering [37–40].

7.2.1 Outlook

The new generation of fluorescence nanoscopy techniques that emerged in the last decade has opened a huge promise for imaging capabilities with nanoscopic resolution. The use of sequential structured illumination enabled more efficient use of the detected photons, which allowed circumvention of the localization limit imposed by the photostability of the fluorescent molecules, resulting in molecular scale spatial resolution. SML-SSI methods have already shown that they can be an invaluable tool to answer biological questions, thanks to unrivaled resolution [13, 15].

MINFLUX has already been commercialized and made available to biologists, although with a steep entry price. The more recent RASTMIN approach might help reducing the entry barrier of SML-SSI measurements, thanks to open-source acquisition and analysis software, open-hardware sample stabilization modules [11, 41], and ease of implementation into laser scanning fluorescence microscopes. We hope this will result in an increased number of these techniques to be routinely used. Just as MINFLUX, RASTMIN can also be extended to 3D and use in an iterative way to obtain increasingly better localization precisions.

However, imaging with SML-SSI methods must be done with caution, and fully exploiting the 1 nm localization precision is non-trivial. Most strategies for fluorescent labeling biological samples introduce significantly larger errors. For example, the distance between fluorophores and target biomolecules can be as large as 10 nm for indirect immunostaining techniques. As a result, the final map of molecule localizations may not directly correspond to the target structure. Fluorophores with improved switching kinetics that are smaller, brighter, and more stable will definitely improve the efficiency of SML-SSI [42–49]. Dyes that can blink spontaneously [50, 51] or can be photoactivated [52] will decouple molecule switching mechanics from the fluorescence excitation, which should further increase the versatility of SML-SSI. A recent combination of MINFLUX with DNA-PAINT [53] showed that the fluorophore photobleaching can become irrelevant while also enabling labeling multiplexing. Use of multiphoton excitation will improve the penetration depth, and it may reduce out-of-focus light and background caused by autofluorescence [54, 55]. Machine or deep learning approaches might improve the throughput of the data analysis and reconstruction, as it was done for camera-based localization methods [56, 57]. SML-SSI methods can also easily benefit from quantitative fluorescence measurements such as fluorescence lifetime (FLIM), as it was already shown in previous works [11, 22], as well as combined with super-resolution fluorescence resonance energy transfer (FRET) [58] or metal or graphene induced energy transfer methods [59–62].

The photostability limitation that motivated the development of SML-SSI has been also tackled by other approaches such as DNA-PAINT, [4, 5] self-healing dyes [63, 64], and photostabilizing imaging buffers [65–67]. Following this trend, it is not out of scope to think that SMLM will soon reach 3 nm localization precision on routine measurements, which is more than enough for current and future labeling strategies and drift correction precisions. Taking this into account, the higher complexity and lower throughput of SML-SSI methods will soon be unjustified. However, there is a new, and very interesting edge for the application of SML-SSI. Since they are significantly more photon efficient than any other method, they could be applied to obtained super-resolution imaging with decent 10–20 nm resolution using "bad" fluorophores, such as fluorescent proteins, something impossible to achieve by any other way.

SML-SSI has also proved to be an excellent tool to help answering biological questions through single molecule tracking. As it requires much fewer photons for reaching required localization precision, fluorophores can be tracked at increased rates and for longer periods before photodegradation. To this end, the more recent pulsed-interleaved MINFLUX (p-MINFLUX) [11] offers the best performance, as the localization measurements can be performed at the repetition rate of the laser in the MHz range.

We expect to see in the coming years an increased number of applications of SML-SSI systems in nanoscopic imaging and ultra-fast single molecule tracking, not only limited to biological sciences. SML-SSI systems can be used in materials science for investigation of nanoporous materials or dynamic synthetic nanostructures (e.g., DNA origami), or thanks to addition of FRET modality, conduct studies of nanophotonic devices, conducting polymers, or multichromophoric systems. Thanks to solid theory and fundamentals behind single molecule localization with sequential structured illumination and established experimental realizations, the near future should bring a wide range of new applications for deeper investigation of both synthetic and natural nanoscopic systems.

REFERENCES

1. M. Lelek, M. T. Gyparaki, G. Beliu, F. Schueder, J. Griffié, S. Manley, R. Jungmann, M. Sauer, M. Lakadamyali, and C. Zimmer, "Single-molecule localization microscopy," *Nat. Rev. Methods Primer* **1**, 1–27 (2021).
2. R. E. Thompson, D. R. Larson, and W. W. Webb, "Precise nanometer localization analysis for individual fluorescent probes," *Biophys. J.* **82**, 2775–2783 (2002).

3. K. I. Mortensen, L. S. Churchman, J. A. Spudich, and H. Flyvbjerg, "Optimized localization analysis for single-molecule tracking and super-resolution microscopy," *Nat. Methods* **7**, 377–381 (2010).

4. J. Schnitzbauer, M. T. Strauss, T. Schlichthaerle, F. Schueder, and R. Jungmann, "Super-resolution microscopy with DNA-PAINT," *Nat. Protoc.* **12**, 1198–1228 (2017).

5. R. Jungmann, C. Steinhauer, M. Scheible, A. Kuzyk, P. Tinnefeld, and F. C. Simmel, "Single-molecule kinetics and super-resolution microscopy by fluorescence imaging of transient binding on DNA origami," *Nano Lett.* **10**, 4756–4761 (2010).

6. M. Dai, R. Jungmann, and P. Yin, "Optical imaging of individual biomolecules in densely packed clusters," *Nat. Nanotechnol.* **11**, 798–807 (2016).

7. S. Strauss, P. C. Nickels, M. T. Strauss, V. Jimenez Sabinina, J. Ellenberg, J. D. Carter, S. Gupta, N. Janjic, and R. Jungmann, "Modified aptamers enable quantitative sub-10-nm cellular DNA-PAINT imaging," *Nat. Methods* **15**, 685–688 (2018).

8. S. Strauss and R. Jungmann, "Up to 100-fold speed-up and multiplexing in optimized DNA-PAINT," *Nat. Methods* **17**, 789–791 (2020).

9. F. Balzarotti, Y. Eilers, K. C. Gwosch, A. H. Gynnå, V. Westphal, F. D. Stefani, J. Elf, and S. W. Hell, "Nanometer resolution imaging and tracking of fluorescent molecules with minimal photon fluxes," *Science* **355**, 606–612 (2017).

10. K. C. Gwosch, J. K. Pape, F. Balzarotti, P. Hoess, J. Ellenberg, J. Ries, and S. W. Hell, "MINFLUX nanoscopy delivers 3D multicolor nanometer resolution in cells," *Nat. Methods* **17**, 217–224 (2020).

11. L. A. Masullo, F. Steiner, J. Zähringer, L. F. Lopez, J. Bohlen, L. Richter, F. Cole, P. Tinnefeld, and F. D. Stefani, "Pulsed Interleaved MINFLUX," *Nano Lett.* **21**, 840–846 (2021).

12. R. Schmidt, T. Weihs, C. A. Wurm, I. Jansen, J. Rehman, S. J. Sahl, and S. W. Hell, "MINFLUX nanometer-scale 3D imaging and microsecond-range tracking on a common fluorescence microscope," *Nat. Commun.* **12**, 1478 (2021).

13. J. O. Wolff, L. Scheiderer, T. Engelhardt, J. Engelhardt, J. Matthias, and S. W. Hell, "MINFLUX dissects the unimpeded walking of kinesin-1," *Science* **379**(6636), 1004–1010 (2023). DOI: 10.1126/science.ade2650

14. Y. Eilers, H. Ta, K. C. Gwosch, F. Balzarotti, and S. W. Hell, "MINFLUX monitors rapid molecular jumps with superior spatiotemporal resolution," *Proc. Natl. Acad. Sci.* **115**, 6117–6122 (2018).

15. T. Deguchi, M. K. Iwanski, E.-M. Schentarra, C. Heidebrecht, L. Schmidt, J. Heck, T. Weihs, S. Schnorrenberg, P. Hoess, S. Liu, V. Chevyreva, K.-M. Noh, L. C. Kapitein, and J. Ries, "Direct observation of motor protein stepping in living cells using MINFLUX," *Science* **379**, 1010–1015 (2022).

16. J. Enderlein, "Tracking of fluorescent molecules diffusing within membranes," *Appl. Phys. B* **71**, 773–777 (2000).

17. M. Weber, M. Leutenegger, S. Stoldt, S. Jakobs, T. S. Mihaila, A. N. Butkevich, and S. W. Hell, "MINSTED fluorescence localization and nanoscopy," *Nat. Photonics* **15**, 361–366 (2021).

18. T. A. Klar, S. Jakobs, M. Dyba, A. Egner, and S. W. Hell, "Fluorescence microscopy with diffraction resolution barrier broken by stimulated emission," *Proc. Natl. Acad. Sci.* **97**, 8206–8210 (2000).

19. T. J. Gould, D. Burke, J. Bewersdorf, and M. J. Booth, "Adaptive optics enables 3D STED microscopy in aberrating specimens," *Opt. Express* **20**, 20998 (2012).

20. P. Zdańkowski, M. Trusiak, D. McGloin, and J. R. Swedlow, "Numerically enhanced stimulated emission depletion microscopy with adaptive optics for deep-tissue super-resolved imaging," *ACS Nano* **14**, 394–405 (2020).

21. L. A. Masullo, L. F. Lopez, and F. D. Stefani, "A common framework for single-molecule localization using sequential structured illumination," *Biophys. Rep.* **2**, 100036 (2022).

22. L. A. Masullo, A. M. Szalai, L. F. Lopez, M. Pilo-Pais, G. P. Acuna, and F. D. Stefani, "An alternative to MINFLUX that enables nanometer resolution in a confocal microscope," *Light Sci. Appl.* **11**, 199 (2022).

23. J. C. Thiele, D. A. Helmerich, N. Oleksiievets, R. Tsukanov, E. Butkevich, M. Sauer, O. Nevskyi, and J. Enderlein, "Confocal fluorescence-lifetime single-molecule localization microscopy," *ACS Nano* **14**, 14190–14200 (2020).

24. Q. Wang and W. E. Moerner, "Optimal strategy for trapping single fluorescent molecules in solution using the ABEL trap," *Appl. Phys. B* **99**, 23–30 (2010).

25. V. Levi, Q. Ruan, K. Kis-Petikova, and E. Gratton, "Scanning FCS, a novel method for three-dimensional particle tracking," *Biochem. Soc. Trans.* **31**, 997–1000 (2003).

26. E. Marklund, B. van Oosten, G. Mao, E. Amselem, K. Kipper, A. Sabantsev, A. Emmerich, D. Globisch, X. Zheng, L. C. Lehmann, O. G. Berg, M. Johansson, J. Elf, and S. Deindl, "DNA surface exploration and operator bypassing during target search," *Nature* **583**, 858–861 (2020).

27. L. M. Davis, B. K. Canfield, J. A. Germann, J. K. King, W. N. Robinson, A. D. Dukes III, S. J. Rosenthal, P. C. Samson, and J. P. Wikswo, "Four-focus single-particle position determination in a confocal microscope," in *Single Molecule Spectroscopy and Imaging III* (SPIE, 2010), Vol. 7571, pp. 140–149.

28. J. A. Germann and L. M. Davis, "Three-dimensional tracking of a single fluorescent nanoparticle using four-focus excitation in a confocal microscope," *Opt. Express* **22**, 5641 (2014).

29. S. M. Kay, *Fundamentals of Statistical Signal Processing: Estimation Theory* (Prentice-Hall, Inc., 1993).

30. J. Chao, E. Sally Ward, and R. J. Ober, "Fisher information theory for parameter estimation in single molecule microscopy: tutorial," *J. Opt. Soc. Am. A* **33**, B36 (2016).

31. R. D. Gill and B. Y. Levit, "Applications of the van Trees inequality: a Bayesian Cramér-Rao bound," *Bernoulli* **1**, 59 (1995).

32. M. O. Lenz, H. G. Sinclair, A. Savell, J. H. Clegg, A. C. N. Brown, D. M. Davis, C. Dunsby, M. A. A. Neil, and P. M. W. French, "3-D stimulated emission depletion microscopy with programmable aberration correction," *J. Biophotonics* **7**, 29–36 (2014).

33. M. Reuss, J. Engelhardt, and S. W. Hell, "Birefringent device converts a standard scanning microscope into a STED microscope that also maps molecular orientation," *Opt. Express* **18**, 1049–1058 (2010).

34. L. A. Masullo, A. Bodén, F. Pennacchietti, G. Coceano, M. Ratz, and I. Testa, "Enhanced photon collection enables four dimensional fluorescence nanoscopy of living systems," *Nat. Commun.* **9**, 3281 (2018).

35. S. W. Hell and M. Nagorni, "4Pi confocal microscopy with alternate interference," *Opt. Lett.* **23**, 1567 (1998).

36. A. M. Szalai, B. Siarry, J. Lukin, D. J. Williamson, N. Unsain, A. Cáceres, M. Pilo-Pais, G. Acuna, D. Refojo, D. M. Owen, S. Simoncelli, and F. D. Stefani, "Three-dimensional total-internal reflection fluorescence nanoscopy with nanometric axial resolution by photometric localization of single molecules," *Nat. Commun.* **12**, 517 (2021).

37. S. R. P. Pavani, M. A. Thompson, J. S. Biteen, S. J. Lord, N. Liu, R. J. Twieg, R. Piestun, and W. E. Moerner, "Three-dimensional, single-molecule fluorescence imaging beyond the diffraction limit by using a double-helix point spread function," *Proc. Natl. Acad. Sci.* **106**, 2995–2999 (2009).

38. Y. Shechtman, S. J. Sahl, A. S. Backer, and W. E. Moerner, "Optimal point spread function design for 3D imaging," *Phys. Rev. Lett.* **113**, 133902 (2014).

39. S. Jia, J. C. Vaughan, and X. Zhuang, "Isotropic three-dimensional super-resolution imaging with a self-bending point spread function," *Nat. Photonics* **8**, 302–306 (2014).

40. L. von Diezmann, Y. Shechtman, and W. E. Moerner, "Three-dimensional localization of single molecules for super-resolution imaging and single-particle tracking," *Chem. Rev.* **117**, 7244–7275 (2017).

41. S. Coelho, J. Baek, M. S. Graus, J. M. Halstead, P. R. Nicovich, K. Feher, H. Gandhi, J. J. Gooding, and K. Gaus, "Ultraprecise single-molecule localization microscopy enables in situ distance measurements in intact cells," *Sci. Adv.* **6**, eaay8271 (2020).

42. S. van de Linde, M. Heilemann, and M. Sauer, "Live-cell super-resolution imaging with synthetic fluorophores," *Annu. Rev. Phys. Chem.* **63**, 519–540 (2012).

43. M. Fernández-Suárez and A. Y. Ting, "Fluorescent probes for super-resolution imaging in living cells," *Nat. Rev. Mol. Cell Biol.* **9**, 929–943 (2008).

44. J. B. Grimm, A. N. Tkachuk, L. Xie, H. Choi, B. Mohar, N. Falco, K. Schaefer, R. Patel, Q. Zheng, Z. Liu, J. Lippincott-Schwartz, T. A. Brown, and L. D. Lavis, "A general method to optimize and functionalize red-shifted rhodamine dyes," *Nat. Methods* **17**, 815–821 (2020).

45. T. S. Mihaila, C. Bäte, L. M. Ostersehlt, J. K. Pape, J. Keller-Findeisen, S. J. Sahl, and S. W. Hell, "Enhanced incorporation of subnanometer tags into cellular proteins for fluorescence nanoscopy via optimized genetic code expansion," *Proc. Natl. Acad. Sci.* **119**, e2201861119 (2022).

46. R. Lincoln, M. L. Bossi, M. Remmel, E. D'Este, A. N. Butkevich, and S. W. Hell, "A general design of caging-group-free photoactivatable fluorophores for live-cell nanoscopy," *Nat. Chem.* **14**, 1013–1020 (2022).

47. L. Wang, M. S. Frei, A. Salim, and K. Johnsson, "Small-molecule fluorescent probes for live-cell super-resolution microscopy," *J. Am. Chem. Soc.* **141**, 2770–2781 (2019).

48. G. Lukinavičius, C. Blaukopf, E. Pershagen, A. Schena, L. Reymond, E. Derivery, M. Gonzalez-Gaitan, E. D'Este, S. W. Hell, D. Wolfram Gerlich, and K. Johnsson, "SiR–Hoechst is a far-red DNA stain for live-cell nanoscopy," *Nat. Commun.* **6**, 8497 (2015).

49. G. Lukinavičius, K. Umezawa, N. Olivier, A. Honigmann, G. Yang, T. Plass, V. Mueller, L. Reymond, I. R. Corrêa Jr, Z.-G. Luo, C. Schultz, E. A. Lemke, P. Heppenstall, C. Eggeling, S. Manley, and K. Johnsson, "A near-infrared fluorophore for live-cell super-resolution microscopy of cellular proteins," *Nat. Chem.* **5**, 132–139 (2013).

50. S. Uno, M. Kamiya, T. Yoshihara, K. Sugawara, K. Okabe, M. C. Tarhan, H. Fujita, T. Funatsu, Y. Okada, S. Tobita, and Y. Urano, "A spontaneously blinking fluorophore based on intramolecular spirocyclization for live-cell super-resolution imaging," *Nat. Chem.* **6**, 681–689 (2014).

51. M. Remmel, L. Scheiderer, A. N. Butkevich, M. L. Bossi, and S. W. Hell, "Spontaneously blinking fluorophores for accelerated MINFLUX nanoscopy," bioRxiv 2022.08.29.505670 (2022).

52. J. B. Grimm, B. P. English, H. Choi, A. K. Muthusamy, B. P. Mehl, P. Dong, T. A. Brown, J. Lippincott-Schwartz, Z. Liu, T. Lionnet, and L. D. Lavis, "Bright photoactivatable fluorophores for single-molecule imaging," *Nat. Methods* **13**, 985–988 (2016).

53. L. M. Ostersehlt, D. C. Jans, A. Wittek, J. Keller-Findeisen, K. Inamdar, S. J. Sahl, S. W. Hell, and S. Jakobs, "DNA-PAINT MINFLUX nanoscopy," *Nat. Methods* **19**, 1072–1075 (2022).

54. L. A. Masullo and F. D. Stefani, "Multiphoton single-molecule localization by sequential excitation with light minima," *Light Sci. Appl.* **11**, 70 (2022).

55. K. Zhao, X. Xu, W. Ren, D. Jin, and P. Xi, "Two-photon MINFLUX with doubled localization precision," *eLight* **2**, 5 (2022).

56. E. Nehme, L. E. Weiss, T. Michaeli, and Y. Shechtman, "Deep-STORM: super-resolution single-molecule microscopy by deep learning," *Optica* **5**, 458–464 (2018).

57. P. Zelger, K. Kaser, B. Rossboth, L. Velas, G. J. Schütz, and A. Jesacher, "Three-dimensional localization microscopy using deep learning," *Opt. Express* **26**, 33166–33179 (2018).

58. A. M. Szalai, C. Zaza, and F. D. Stefani, "Super-resolution FRET measurements," *Nanoscale* **13**, 18421–18433 (2021).

59. A. Ghosh, A. Sharma, A. I. Chizhik, S. Isbaner, D. Ruhlandt, R. Tsukanov, I. Gregor, N. Karedla, and J. Enderlein, "Graphene-based metal-induced energy transfer for sub-nanometre optical localization," *Nat. Photonics* **13**, 860–865 (2019).

60. A. Ghosh, A. I. Chizhik, N. Karedla, and J. Enderlein, "Graphene- and metal-induced energy transfer for single-molecule imaging and live-cell nanoscopy with (sub)-nanometer axial resolution," *Nat. Protoc.* **16**, 3695–3715 (2021).

61. I. Kaminska, J. Bohlen, S. Rocchetti, F. Selbach, G. P. Acuna, and P. Tinnefeld, "Distance dependence of single-molecule energy transfer to graphene measured with DNA origami nanopositioners," *Nano Lett.* **19**, 4257–4262 (2019).

62. I. Kamińska, J. Bohlen, R. Yaadav, P. Schüler, M. Raab, T. Schröder, J. Zähringer, K. Zielonka, S. Krause, and P. Tinnefeld, "Graphene energy transfer for single-molecule biophysics, biosensing, and super-resolution microscopy," *Adv. Mater.* **33**, 2101099 (2021).

63. R. B. Altman, Q. Zheng, Z. Zhou, D. S. Terry, J. D. Warren, and S. C. Blanchard, "Enhanced photostability of cyanine fluorophores across the visible spectrum," *Nat. Methods* **9**, 428–429 (2012).

64. M. Isselstein, L. Zhang, V. Glembockyte, O. Brix, G. Cosa, P. Tinnefeld, and T. Cordes, "Self-healing dyes—keeping the promise?," *J. Phys. Chem. Lett.* **11**, 4462–4480 (2020).

65. I. Rasnik, S. A. McKinney, and T. Ha, "Nonblinking and long-lasting single-molecule fluorescence imaging," *Nat. Methods* **3**, 891–893 (2006).

66. C. E. Aitken, R. A. Marshall, and J. D. Puglisi, "An oxygen scavenging system for improvement of dye stability in single-molecule fluorescence experiments," *Biophys. J.* **94**, 1826–1835 (2008).

67. T. Ha and P. Tinnefeld, "Photophysics of fluorescent probes for single-molecule biophysics and super-resolution imaging," *Annu. Rev. Phys. Chem.* **63**, 595–617 (2012).

CHAPTER 8

Measuring Molecule Numbers in Nano-Scale Assemblies With Single-Molecule Localization Microscopy

Marina S. Dietz, Hans-Dieter Barth, and Mike Heilemann

Goethe-University Frankfurt, Frankfurt, Germany

8.1 INTRODUCTION

Fluorescence microscopy, a special form of light microscopy, is subject to the limitations of diffraction, resulting in a resolution limit of about 200 nm in the visible light range. Single-molecule localization microscopy (SMLM) is an imaging technique that circumvents this limit and can produce super-resolved images with near-molecular spatial resolution. This significant gain in resolution is achieved by spatially and temporally separating the emission events of individual fluorophores [1, 2]. One way to achieve this separation is to use photoswitchable fluorophores and to adjust the fraction of active fluorophores to a low density, resulting in separate blinking of individual fluorophores. Examples of such fluorophores are photoactivatable or photoconvertible fluorescent proteins, as used in (fluorescence) photoactivated localization microscopy ((F)PALM) [3, 4], or photoswitchable organic fluorophores, as used in (*direct*) stochastic

DOI: 10.1201/9781003220688-8

optical reconstruction microscopy ((d)STORM) [5, 6]. Fluorescence emission from individual fluorophores is first detected as a point spread function (PSF), and subsequently its centroid is determined, for example, by approximation with a Gaussian function. The precision with which the center of the PSF can be determined scales inversely with the square root of the number of detected photons originating from the fluorophore [7]. Finally, a super-resolved image is reconstructed from the collected coordinates of a sufficiently large number of individual fluorophores. The resulting list of coordinates can then be used for post-processing of the imaging data (Figure 8.1).

Although the spatial resolution in SMLM is significantly improved compared to diffraction-limited microscopy, the organization of biomolecules within small assemblies cannot yet be resolved. However, it is precisely the organization at the molecular level that affects the structure and properties of biomaterials, and accordingly it is of central importance to understand these molecular assemblies in order to develop improved materials (e.g., for biomedical applications [8, 9]). SMLM localization data contains more information than the simple spatial reconstruction of the image. Assessing the sample "one molecule at the time" has the advantage of providing information about the number of molecules present in a specific structure (e.g., a nanoparticles or a biological protein assembly), i.e., SMLM acts as a molecular counter. Therefore, SMLM can be expanded to quantitative SMLM (qSMLM) and report on molecular numbers in small assemblies [10]. This chapter presents the quantitative analysis of PALM and STORM data to extract molecular numbers from the blink properties of fluorescent proteins or organic dyes (Figure 8.1d).

8.2 EXTRACTING MOLECULE NUMBERS FROM SINGLE-MOLECULE PALM DATA

8.2.1 Photophysics of Fluorescent Proteins

Quantitative PALM (qPALM) uses photoconvertible or photoactivatable fluorescent proteins whose fluorescence blinking kinetics were investigated by several groups [11–16]. In brief, the fluorescence kinetics can be modeled by defining molecular states (e.g., dark, bright, bleached) and the kinetic constants for the transitions between these states. For the purpose of quantitative SMLM, we will restrict to a four-state model which is simple but sufficient to describe the photophysical transitions in photoactivatable or photoconvertible fluorescent proteins (Figure 8.2). At the beginning of the experiment, all fluorescent proteins are in an undetected

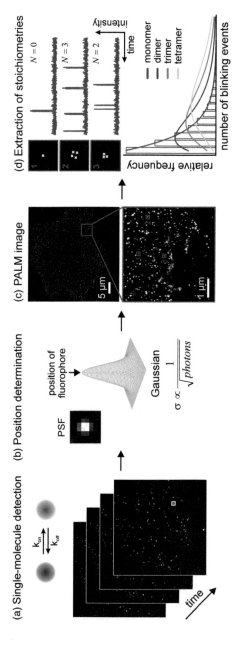

FIGURE 8.1   Principle of quantitative single-molecule localization microscopy: (a) In SMLM, spatially overlapping signals are separated in time by photoswitching fluorophores between a detectable fluorescent and an undetectable non-fluorescent state. A series of fluorescence images with low-density single-molecule emission events is recorded. (b) The PSF of single molecules is approximated with a Gaussian function to obtain the position of the fluorophore. The precision with which the position is determined is anti-proportional to the number of photons detected in the local image spot of a single fluorophore. A super-resolved image is reconstructed from all single-molecule coordinates. (c) PALM image of a HeLa cell showing super-resolved receptor assemblies in the plasma membrane. (d) The number of proteins within each protein assembly (green boxes 1–3 in (c)) can be determined by extracting the number of fluorescence emission events (white crosses), histogramming the number of blinking events N (reoccurrence of fluorescence) for many localization clusters, and approximating this distribution with a model function.

PALM

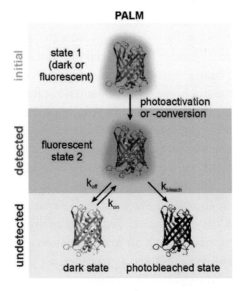

FIGURE 8.2   Four-state model of the photophysical transitions in fluorescent proteins. Photoactivatable fluorescent proteins initially exist in a non-fluorescent state. Photoconvertible fluorescent proteins initially exist in a fluorescent state, however the fluorescence emission from this initial state is not registered in the experiment. In both cases, irradiation with violet light leads to photoactivation/photoconversion of the protein and generates a (new) fluorescent state, the fluorescence of which is now detected in the experiment. From this state, the protein can either enter a transient dark state or irreversibly photobleach.

state (dark or fluorescent in a non-detected channel). When illuminated with low intensities of violet light, stochastic photoactivation/photoconversion into a fluorescent state occurs, which is detected as an emission event. From this fluorescent state, a transient dark state can be reversibly populated. The repopulation of the fluorescent state out of this dark state is counted as the first blinking event. Hence, a fluorescent protein that only produces one emission event, shows no blinking, while a protein exhibiting three emission events shows 2 blinking events (Figure 8.1d, top). Finally, the fluorescent protein is irreversibly destroyed by photobleaching.

8.2.2 Mathematical Models to Describe Blinking Statistics

One important application of qSMLM is the analysis of molecular stoichiometry in complexes. This is for example crucial in biology where proteins can be found as monomers, dimers, or higher order oligomers, each of these associated to a specific biological function. Stoichiometry analyses are also important in synthetic materials and biomaterials that are generally decorated with active ligands. The material functionality strictly depends on the amount of ligand; therefore, the analysis of the stoichiometry is also crucial. For the analysis of the oligomeric state of molecular complexes with qPALM, the number N of blinking events from a sufficiently large number of localization clusters is determined and histogrammed (Figure 8.1d). Herein, a localization cluster is a collection of spatially close localizations originating either from a single fluorescent protein (monomer) or from multiple fluorescent proteins (oligomer). A crucial assumption is that distinct protein complexes are sufficiently separated from each other to be resolved by SMLM to be analyzed individually. Failing to spatially distinguish clusters will result in an overestimation of the stoichiometry. Depending on the oligomeric state, a characteristic distribution of blinking events is produced (Figure 8.1d, bottom).

In the case of a single fluorescent protein (monomer), this histogram is described by a geometric distribution, a distribution with a single parameter p reporting the probability that the fluorophore does not photobleach (Eq. 8.1).

$$P_0(N) = p(1-p)^N \qquad (8.1)$$

The photobleaching probability p is a fluorophore-specific property and is determined experimentally. This can be achieved with a single-molecule surface of the fluorescent protein, or in a cell expressing a monomeric membrane protein fused to the fluorescent protein. p values for fluorescent proteins mostly vary between 0.1 and 0.8 [16–18]. Note that $1/p$ yields the average number of emission events before photobleaching.

For oligomers, the kinetic model is extended by a binomial term with a second parameter q that corrects for incomplete detection of multiple fluorophores within the same cluster. Here, $(1-q)$ corresponds to the detectability of the fluorophore. The general kinetic model for $(m+1)$-mers is shown in Eq. 8.2.

$$P_m(N) = \sum_{k=0}^{\min(m,N)} \frac{m}{k} \frac{N}{k} q^{m-k} (1-q)^k p^{k+1} (1-p)^{N-k} \qquad (8.2)$$

To verify the applicability of the model, blinking histograms can be generated from experimental data of a monomeric fluorescent protein, such that q equals 0, and the detectability equals 1. Oligomers consisting of different molecule numbers are obtained by synthetically grouping the localization clusters of single fluorescent proteins detected on a surface. Subsequently, the blink histograms are approximated with the respective kinetic functions derived for dimers, trimers, and tetramers (Figure 8.3). As can be seen, the model precisely approximates the distribution of the blinking events.

If applied to multi-protein assemblies, the detection of fluorescent proteins is incomplete, and the detectability must be determined experimentally. This can be achieved by measuring the detectability for a reference sample with a known stoichiometry. One option is to generate dimers of fluorescent proteins by attaching them to a synthetic scaffold (e.g., double-stranded DNA). In the example shown in Figure 8.4a the DNA carries two Ni^{2+}-NTA residues which bind with high affinity to His_6-tagged proteins (here mEos3.2). After purification via size exclusion chromatography, the synthetic dimer can be deposited on a surface and a blinking analysis can be performed. For application in cells, reference proteins with a known oligomeric state fused to a fluorescent protein can be used, for example, the dimeric CTLA4 (Figure 8.4b). This calibration is critical for the accurate assessment of the stoichiometry and sources of error such as incorrect protein folding of the standard, background noise, and impurities must be minimized. Moreover it has to be taken in account that, especially for synthetic materials, the chemical environment may affect the blinking kinetics and therefore the standard calibration, and the sample has to be measured as much as possible in the same conditions.

With known photobleaching probability p and detectability $(1-q)$, the oligomeric state of an unknown target can be investigated this way. It is also possible to retrieve information on mixed oligomer populations by using linear combinations of the corresponding kinetic equations. Equation 8.3 describes the blinking histogram of a combination of monomers and dimers where f is the fraction of monomers.

$$P_{0,1}(N) = fp(1-p)^N + (1-f)p(1-p)^{N-1}(Np(1-q)+(1-p)q) \quad (8.3)$$

qPALM reports on the oligomeric state of a large number of protein assemblies but cannot provide this information for a single cluster. In addition,

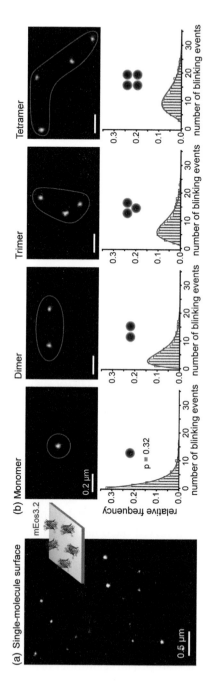

FIGURE 8.3 qPALM measurements of a single-molecule surface of fluorescent proteins. (a) The fluorescent protein mEos3.2 was dispersed on a poly-L-lysine coated glass slide and analyzed with PALM. The super-resolved image shows well-separated localization clusters where each cluster represents a single fluorescent protein. (b) Counting the blinking events of many individual clusters yields a distribution that is well approximated by the geometric distribution (Eq. 8.8.1) describing a monomeric state (left; $p = 0.32$). Histograms for dimers, trimers, and tetramers were generated by counting the blinking events of two, three, or four clusters, respectively. These distributions are described by Eq. 8.8.2, where the p value was determined from the geometric distribution and q is equal to 0 because there are no undetected molecules in these composed oligomers. (Adapted from Baldering et al. Copyright CC BY 4.0.)

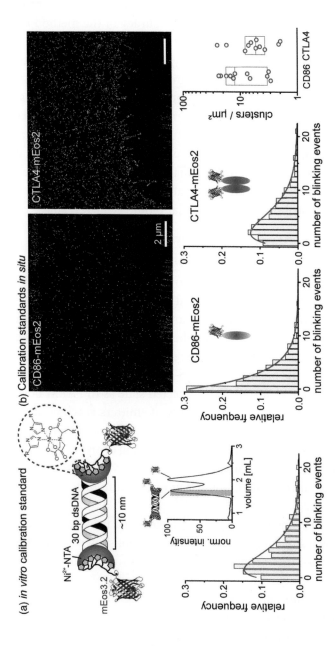

FIGURE 8.4 Reference standards for qPALM *in vitro* and *in situ*. (a) A synthetic dimer of two His$_6$-tagged fluorescent proteins was generated using Ni^{2+}-NTA-functionalized dsDNA. The high-affinity interaction of Ni^{2+}-NTA and His$_6$ allows purification of the complex by size exclusion chromatography and subsequent deposition on a single-molecule surface coated with poly-L-lysine for qPALM analysis. (b) In cells, reference proteins with known monomeric (CD86) and dimeric (CTLA4) stoichiometry are used to determine p and q values. The PALM images show a homogeneous distribution of the two proteins on the plasma membrane of cells. Next to the oligomeric state of protein assemblies, super-resolution microscopy also reports on the cluster density. (Adapted from Fricke et al [14]. Copyright CC BY 4.0.)

the qPALM analysis assumes the detection of at least one fluorophore within a cluster [16] and neglects clusters with no active fluorophores. Baldering et al. have extended the analysis and included the presence of non-fluorescent clusters, which improves the accuracy of the method for larger oligomers [18].

In addition to determining stoichiometry, other information can be obtained from SMLM data. Counting biomolecules on a surface (e.g., receptors on the plasma membrane of cells), provides information on cluster densities (Figure 8.4b, bottom right). From this, if both stoichiometry and cluster density are known, this can be used to determine the total number of molecules in a sample and to formulate chemical equations that describe functional transitions. In further analyses, the distances between clusters can be analyzed, providing information on the homogeneity or heterogeneity of the distribution of a particular biomolecule [19, 20].

8.3 EXTRACTING MOLECULE NUMBERS FROM SINGLE-MOLECULE *D*STORM DATA

Kinetics-based quantitative analysis can also be applied to photoswitchable organic fluorophores, such as those used in *d*STORM. In contrast to fluorescent proteins, Alexa Fluor 647 [21], an organic dye often used in *d*STORM experiments, cycles several times between an ON and OFF state before photobleaching. The main difference of qPALM with respect to the kinetic model is that at the beginning of a *d*STORM experiment all emitters are in the ON state (i.e., an initial undetected state as in PALM does not exist) (see also Figure 8.2). The low density of emitters is achieved by light-induced transfer of the majority of fluorophores into the OFF state. Due to the high density of fluorophores at the beginning of an experiment, the first emission event is not captured in the quantitative analysis, which can be corrected for in the kinetic modeling of blinking histograms of organic fluorophores [22].

As in qPALM, the parameters p and q are determined from calibration standards with known stoichiometry. In the experiment shown below, the parameter p was determined from a single-molecule surface functionalized with biotin and streptavidin that binds a short dsDNA carrying a single Alexa Fluor 647 fluorophore. Here, a value of 0.054 was obtained for p. The parameter q was determined from a DNA origami containing three fluorophores in each of two corners, with a corner-to-corner distance of 120 nm (Figure 8.5a). The two trimers of Alexa Fluor 647 are resolved as contiguous spots by *d*STORM, while the individual fluorophores within a

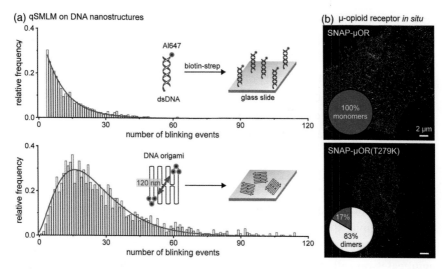

FIGURE 8.5 Quantitative *d*STORM measurements. (a) DNA nanostructures with a monomer or a trimer stoichiometry were used to validate the application of *d*STORM quantification. Glass slides were modified with biotin and streptavidin and either single-labeled dsDNA or DNA origami with two groups of Alexa Fluor 647 (Al647) trimers were immobilized on this functionalized surface. The histograms of the blinking events yielded the expected distributions for monomers and trimers. (b) Stoichiometry analysis of the μ-opioid receptor (μOR) *in situ*. SNAP-tagged receptors were labeled with Alexa Fluor 647 in CHO-K1 cells and analyzed with *d*STORM. The distribution of blinking events yielded for the untreated wild-type μOR a pure monomer, while the constitutively active variant T279K occurs predominantly as a dimer. (Data taken from Möller et al. [23].)

trimer were not resolved. This design allowed distinguishing trimers from existing background signals. The analysis yielded that essentially all fluorophores were detected and thus the parameter q approaches the value of 0. In addition, it was found that Alexa Fluor 647 is sensitive to the nano-environment, as slightly different photobleaching probabilities p were found for the three dyes in a trimer. Different nucleobases in the immediate vicinity of the dyes can, for example, hinder their rotation or lead to nano-environment specific fluorescence quenching. DNA origami are a powerful tool to generate standards with controlled distances and stoichiometry, however, note that they may also bring some errors due to missing strands during the folding process that decrease stoichiometry.

In a study according to the principle described above, quantitative *d*STORM was used to determine the oligomeric state of the G-protein coupled μ-opioid receptor (μOR) [23] (Figure 8.5b). A monomeric and a dimeric reference protein, β_1-adrenergic receptor and CD28, were used to determine p and q for Alexa Fluor 647 on the plasma membrane of cells. The wild-type μOR was found to be monomeric in untreated cells, while the constitutively active T279K variant shows a large proportion of dimers. Stoichiometric fluorophore labeling of the membrane receptors was achieved through the SNAP-tag [24], which embeds the fluorophore inside the protein tag and ensures a nearly identical nano-environment.

8.4 DISCUSSION

Analysis of biomaterials at the molecular level is essential for understanding their structure and function. In addition to the spatial arrangement of, for example, structural proteins, it is also of great interest to study the behavior as well as the organization of receptors, which often occur in aggregates on cell surfaces and cannot be clearly resolved even by super-resolution methods. Although the molecular components within small protein assemblies cannot be visualized directly with SMLM techniques, it is possible to retrieve information on molecule numbers by analyzing the available kinetic information inherent to SMLM data. The origin of this kinetic information in PALM and dSTORM are reversible transitions of fluorophores between fluorescent and dark states. The corresponding blinking behavior resulting from switching between these states contains valuable information and is used as input by several methods that are capable of reporting numbers of molecules in small assemblies [25, 26]. The detected blinking events can be modeled with kinetic equations which themselves are linked to molecule numbers. However, it is important to keep in mind that the photophysical transitions of fluorophores are influenced by their nano-environment. For example, organic fluorophores are "un-shielded" and their blinking statistics are affected by local conditions (e.g., the molecule the fluorophore is attached to) and the global chemical environment (e.g., redox-active reagents in a buffer) [22, 27]. Ideally, organic dyes should be positioned in a defined nano-environment to obtain reliable stoichiometries, by using for example protein tags such as SNAP- and HALO-tags which shield the fluorophore [24, 28]. The blinking statistics of fluorescent proteins with their chromophore embedded in a β-barrel structure are modulated to a lesser extent in the presence of oxidizing or reducing agents [18, 29]. Of great importance in the analysis of blinking statistics for green-to-orange photoconvertible fluorescent proteins, such as variants from the mEos family, are recently discovered photophysics of the green state and the occurrence of long OFF states, which might affect the accuracy of the obtained results [30–33]. The analysis of blinking statistics in quantitative PALM was facilitated by an automated quantitative approach for fluorescence kinetics analysis (QAFKA), which combines single-molecule detection and stoichiometry analysis [34].

Other alternative SMLM methods not presented in this article generate a single-molecule signal using weak-affinity labels that transiently and repetitively bind to a target. This can for example be achieved by the hybridization of short DNA oligonucleotides, as in DNA-PAINT where a fluorescently labeled strand couples to a structurally bound complementary strand [35]. Since these association and dissociation events are also governed by kinetic equations, they offer access to molecular quantification, as in quantitative PAINT (qPAINT) [36]. An advantage of DNA-PAINT is that recurrent binding events of fluorophore labels to a target yield a large number of emission events per single target, which allows the extraction of the oligomeric state from even a single cluster. Furthermore, by adjusting the fluorophore concentration in the buffer, qPAINT covers a wider dynamic range. qPAINT, on the other hand, necessitates an extra labeling step, such as a DNA-labeled antibody targeting a protein of interest and a fluorophore-labeled single-stranded DNA targeting the antibody. This may result in non-stoichiometric labeling and/or unspecific detection events, which are mostly avoided when utilizing fluorescent proteins or protein tags. Future developments will aim to combine the strengths of both approaches (i.e., genetic and stoichiometric labeling of a target as well as a single-cluster readout).

REFERENCES

1. M. Sauer and M. Heilemann, "Single-molecule localization microscopy in eukaryotes," *Chem. Rev.* **117**, 7478–7509 (2017).
2. M. Lelek, M. T. Gyparaki, G. Beliu, F. Schueder, J. Griffié, S. Manley, R. Jungmann, M. Sauer, M. Lakadamyali, and C. Zimmer, "Single-molecule localization microscopy," *Nat. Rev. Methods Primers* **1**, 1–27 (2021).
3. E. Betzig, G. H. Patterson, R. Sougrat, O. W. Lindwasser, S. Olenych, J. S. Bonifacino, M. W. Davidson, J. Lippincott-Schwartz, and H. F. Hess, "Imaging intracellular fluorescent proteins at nanometer resolution," *Science* **313**, 1642–1645 (2006).
4. S. T. Hess, T. P. K. Girirajan, and M. D. Mason, "Ultra-high resolution imaging by fluorescence photoactivation localization microscopy," *Biophys. J.* **91**, 4258–4272 (2006).
5. M. J. Rust, M. Bates, and X. Zhuang, "Sub-diffraction-limit imaging by stochastic optical reconstruction microscopy (STORM)," *Nat. Methods* **3**, 793–795 (2006).

6. M. Heilemann, S. van de Linde, M. Schüttpelz, R. Kasper, B. Seefeldt, A. Mukherjee, P. Tinnefeld, and M. Sauer, "Subdiffraction-resolution fluorescence imaging with conventional fluorescent probes," *Angew. Chem. Int. Ed.* **47**, 6172–6176 (2008).

7. K. I. Mortensen, L. S. Churchman, J. A. Spudich, and H. Flyvbjerg, "Optimized localization analysis for single-molecule tracking and super-resolution microscopy," *Nat. Methods* **7**, 377–381 (2010).

8. S. Pujals and L. Albertazzi, "Super-resolution microscopy for nanomedicine research," *ACS Nano* **13**, 9707–9712 (2019).

9. S. Pujals, N. Feiner-Gracia, P. Delcanale, I. Voets, and L. Albertazzi, "Super-resolution microscopy as a powerful tool to study complex synthetic materials," *Nat. Rev. Chem.* **3**, 68–84 (2019).

10. M. S. Dietz and M. Heilemann, "Optical super-resolution microscopy unravels the molecular composition of functional protein complexes," *Nanoscale* **11**, 17981–17991 (2019).

11. P. Annibale, S. Vanni, M. Scarselli, U. Rothlisberger, and A. Radenovic, "Quantitative photo activated localization microscopy: unraveling the effects of photoblinking," *PLOS ONE* **6**, e22678 (2011).

12. S.-H. Lee, J. Y. Shin, A. Lee, and C. Bustamante, "Counting single photo-activatable fluorescent molecules by photoactivated localization microscopy (PALM)," *Proc. Nat. Acad. Sci.* **109**, 17436–17441 (2012).

13. S. Avilov, R. Berardozzi, M. S. Gunewardene, V. Adam, S. T. Hess, and D. Bourgeois, "In cellulo evaluation of phototransformation quantum yields in fluorescent proteins used as markers for single-molecule localization microscopy," *PLOS ONE* **9**, e98362 (2014).

14. F. Fricke, J. Beaudouin, R. Eils, and M. Heilemann, "One, two or three? Probing the stoichiometry of membrane proteins by single-molecule localization microscopy," *Sci. Rep.* **5**, 14072 (2015).

15. G. C. Rollins, J. Y. Shin, C. Bustamante, and S. Pressé, "Stochastic approach to the molecular counting problem in superresolution microscopy," *PNAS* **112**, E110–E118 (2015).

16. G. Hummer, F. Fricke, and M. Heilemann, "Model-independent counting of molecules in single-molecule localization microscopy," *Mol. Biol. Cell* **27**, 3637–3644 (2016).

17. C. L. Krüger, M.-T. Zeuner, G. S. Cottrell, D. Widera, and M. Heilemann, "Quantitative single-molecule imaging of TLR4 reveals ligand-specific receptor dimerization," *Sci. Signal* **10**, eaan1308 (2017).

18. T. N. Baldering, J. T. Bullerjahn, G. Hummer, M. Heilemann, and S. Malkusch, "Molecule counts in complex oligomers with single-molecule localization microscopy," *J. Phys. D: Appl. Phys.* **52**, 474002 (2019).

19. M. S. Schröder, M.-L. I. E. Harwardt, J. V. Rahm, Y. Li, P. Freund, M. S. Dietz, and M. Heilemann, "Imaging the fibroblast growth factor receptor network on the plasma membrane with DNA-assisted single-molecule super-resolution microscopy," *Methods* **193**, 38–45 (2021).

20. L. S. Fischer, C. Klingner, T. Schlichthaerle, M. T. Strauss, R. Böttcher, R. Fässler, R. Jungmann, and C. Grashoff, "Quantitative single-protein imaging reveals molecular complex formation of integrin, talin, and kindlin during cell adhesion," *Nat. Commun.* **12**, 919 (2021).

21. M. Heilemann, E. Margeat, R. Kasper, M. Sauer, and P. Tinnefeld, "Carbocyanine dyes as efficient reversible single-molecule optical switch," *J. Am. Chem. Soc.* **127**, 3801–3806 (2005).

22. C. Karathanasis, F. Fricke, G. Hummer, and M. Heilemann, "Molecule counts in localization microscopy with organic fluorophores," *ChemPhysChem* **18**, 942–948 (2017).

23. J. Möller, A. Isbilir, T. Sungkaworn, B. Osberg, C. Karathanasis, V. Sunkara, E. O. Grushevskyi, A. Bock, P. Annibale, M. Heilemann, C. Schütte, and M. J. Lohse, "Single-molecule analysis reveals agonist-specific dimer formation of μ-opioid receptors," *Nat. Chem. Biol.* **16**, 946–954 (2020).

24. A. Keppler, S. Gendreizig, T. Gronemeyer, H. Pick, H. Vogel, and K. Johnsson, "A general method for the covalent labeling of fusion proteins with small molecules in vivo," *Nat. Biotechnol.* **21**, 86–89 (2003).

25. S.-H. Lee, J. Y. Shin, A. Lee, and C. Bustamante, "Counting single photo-activatable fluorescent molecules by photoactivated localization microscopy (PALM)," *PNAS* **109**, 17436–17441 (2012).

26. F. C. Zanacchi, C. Manzo, A. S. Alvarez, N. D. Derr, M. F. Garcia-Parajo, and M. Lakadamyali, "A DNA origami platform for quantifying protein copy number in super-resolution," *Nat. Method* advance online publication, (2017).

27. S. Nanguneri, B. Flottmann, F. Herrmannsdörfer, K. Thomas, and M. Heilemann, "Single-molecule super-resolution imaging by tryptophan-quenching-induced photoswitching of phalloidin-fluorophore conjugates," *Microsc. Res. Tech.* **77**, 510–516 (2014).

28. G. V. Los, L. P. Encell, M. G. McDougall, D. D. Hartzell, N. Karassina, C. Zimprich, M. G. Wood, R. Learish, R. F. Ohana, M. Urh, D. Simpson, J. Mendez, K. Zimmerman, P. Otto, G. Vidugiris, J. Zhu, A. Darzins, D. H. Klaubert, R. F. Bulleit, and K. V. Wood, "HaloTag: a novel protein labeling technology for cell imaging and protein analysis," *ACS Chem. Biol.* **3**, 373–382 (2008).

29. T. N. Baldering, M. S. Dietz, K. Gatterdam, C. Karathanasis, R. Wieneke, R. Tampé, and M. Heilemann, "Synthetic and genetic dimers as quantification ruler for single-molecule counting with PALM," *MBoC* **30**, 1369–1376 (2019).

30. K. Nienhaus and G. U. Nienhaus, "Fluorescent proteins of the EosFP clade: intriguing marker tools with multiple photoactivation modes for advanced microscopy," *RSC Chem. Biol.* **2**, 796–814 (2021).

31. D. Thédié, R. Berardozzi, V. Adam, and D. Bourgeois, "Photoswitching of green mEos2 by intense 561 nm light perturbs efficient green-to-red photoconversion in localization microscopy," *J. Phys. Chem. Lett.* **8**, 4424–4430 (2017).

32. M. Sun, K. Hu, J. Bewersdorf, and T. D. Pollard, "Sample preparation and imaging conditions affect mEos3.2 photophysics in fission yeast cells," *Biophys. J.* **120**, 21–34 (2021).

33. E. De Zitter, J. Ridard, D. Thédié, V. Adam, B. Lévy, M. Byrdin, G. Gotthard, L. Van Meervelt, P. Dedecker, I. Demachy, and D. Bourgeois, "Mechanistic investigations of green mEos4b reveal a dynamic long-lived dark state," *J. Am. Chem. Soc.* **142**, 10978–10988 (2020).

34. A. Saguy, T. N. Baldering, L. E. Weiss, E. Nehme, C. Karathanasis, M. S. Dietz, M. Heilemann, and Y. Shechtman, "Automated analysis of fluorescence kinetics in single-molecule localization microscopy data reveals protein stoichiometry," *J. Phys. Chem. B* **125**, 5716–5721 (2021).

35. R. Jungmann, C. Steinhauer, M. Scheible, A. Kuzyk, P. Tinnefeld, and F. C. Simmel, "Single-molecule kinetics and super-resolution microscopy by fluorescence imaging of transient binding on DNA origami," *Nano Lett.* **10**, 4756–4761 (2010).

36. R. Jungmann, M. S. Avendaño, M. Dai, J. B. Woehrstein, S. S. Agasti, Z. Feiger, A. Rodal, and P. Yin, "Quantitative super-resolution imaging with qPAINT," *Nat. Method* advance online publication, (2016).

CHAPTER 9

Super-Resolution Microscopy in Colloid Science

Ilja Voets

Eindhoven University of Technology, Eindhoven, Netherlands

Frank Scheffold

University of Fribourg, Fribourg, Switzerland

9.1 INTRODUCTION

The diversity in composition, dimensions, and functionality make colloids indispensable building blocks of most present-day materials. Consequentially, colloids transcend virtually all facets of our daily lives, from the milk in our breakfast cereal and tooth paste we brush our teeth with, to the paints on our home and office walls and the protective polymer coatings on our cars, windows, and mobile phones.

But what are 'colloids'? Briefly, colloids or colloidal systems are loosely defined as heterogeneous systems comprising two phases, one dispersed in another, with at least one dimension within the colloidal domain of a few nanometers to a few micrometers. The colloidal domain is thus bracketed by atomic and molecular materials on the one hand and granular matter on the other (Figure 9.1). Colloidal systems are further classified through various means, commonly according to the nature of the dispersed and continuous phases. Emulsions, for example, comprise two immiscible liquids, one typically dispersed as droplets in the other. (Solid) aerosols are another well-known example, consisting for example of airborne viral particles in a sigh of breath.

DOI: 10.1201/9781003220688-9

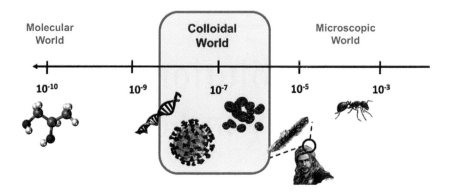

FIGURE 9.1 Colloidal domain. (Adapted with permission from [1].)

Many colloidal systems contain colloidal particles. These solid-like building blocks are often composed of polymers or inorganic materials, such as the latex and titanium dioxide particles used as fillers and pigments of many water-borne paints. Colloidal particles are the dispersed phase of suspensions and may also be used as stabilizers of interfaces in other colloids, such as emulsions and foams. Interface stabilization by surfactants, polymers and/or particles is of critical importance for the colloidal stability and shelf-life of colloids. Without stabilizers that generate a steric and/or electrostatic barrier against (ir)reversible aggregation and other breakdown processes, the internal structure of colloids coarsens over time so as to reduce the total amount of interfacial area that is present in the materials. Even modest energy costs per nm^2 of interface are sufficient to drive such instabilities, since colloids contain an incredibly large interfacial area. For example, a 1 L of a 1% suspension of beads with a radius of 5 nm contains approximately 1 soccer field of total interfacial area.

One might argue that colloidal interfaces are as important as challenging to characterize. Whilst interfaces are often assumed to be homogeneous (i.e., in most theoretical descriptions of colloidal particles), they rarely are in reality. Information on both intra- and interparticle heterogeneities that may be critical for particle functionality is typically lost in ensemble measurements of surface composition and charge. Moreover, the mass of material in the interfacial region compared to the particle bulk is very small (in particular for large, micron-sized colloids). Particle characterization methods relying on conventional tools that are not surface-sensitive, therefore, report mostly on bulk rather than on interfacial properties. Tools that are interface-specific often lack the required lateral resolution to detect (nanoscopic) intraparticle heterogeneities or offer qualitative rather than quantitative information. Recently, super-resolution microscopy (SMLM) tools have been employed to address these challenges in the characterization of colloidal particles aiming ultimately to elucidate structure-activity relations in more detail. This would create a knowledge base, which is essential to tailor and optimize particle synthesis protocols toward reproducible manufacturing procedures to prepare particles with a well-characterized and custom-tailored interface to fine-tune colloidal stability and boost the performance of products and devices containing such particles.

In this chapter, we will highlight recent studies on particle-laden interfaces and microgels wherein super-resolution microscopy has been applied to take a closer look at the relation between the properties of the individual particles and their collective behavior at interfaces and in bulk.

9.2 COLLOIDS AT INTERFACES

Traditionally, interfaces in colloids are stabilized by surfactants or polymers. More recently however, particles have been used for the same purpose. Emulsions stabilized in this manner became known as Pickering emulsions. Particularly appealing of this strategy is the high degree of stability that can be achieved with a wide variety of particles types, such that emulsions can be kept far longer than usual and without the risk of destabilization of active ingredients, such as proteins or cells, due to close contact with denaturing surfactants.

While the relation between particle properties, such as size, shape, surface composition, charge and roughness, and emulsion stabilization have been widely investigated since the introduction of Pickering emulsions, many fundamental questions on interface stabilization with (sub) micron-sized particles remain. In particular, the impact of (variations in) particle properties and local interface stabilization and non-equilibrium phenomena is not yet clear. This is to a large extent due to the inaccessibility of direct, particle-resolved experimental information on the position of small, colloidal particles at deformable interfaces. Various strategies were developed in the past decade to address this challenge, and with success. These shed light on the impact of, for example, particle roughness on interface pinning [2, 3] and microgel structure on their interfacial morphology [4]. Disadvantageously, many of these tools require complex infrastructure and/or involve a number of complex sample preparation steps, such as gelation, vitrification, and shadow-casting.

In a series of recent papers, super-resolution microscopy was introduced as an alternative strategy to image *in-situ* in a minimally invasive manner interface-bound colloidal particles to relate (in a particle-resolved manner) the properties of particles to their contact angle at deformable interfaces between two immiscible liquids [5–7]. A first study introduced the methodology for the simultaneous, *in-situ* visualization of both the colloidal particles and the reconstruction of the fluid interface to which they are adsorbed [5]. A newly developed super-resolution microscopy method coined iPAINT [8] was used to stain both the particles and liquid-liquid interface in an aspecific, non-covalent manner with photo-activatable fluorescent probes (Figure 9.2a). Such staining results in temporal trapping of the fluorescent probe at the interface, creating fluorescence bursts that can be localized using techniques described in Chapter 2. Acquisition of a large number of interfacial binding events enables the reconstruction of the geometry of the interface with a precision of several tens of nanometers.

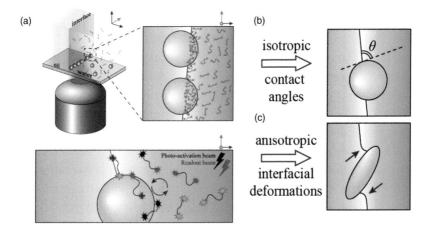

FIGURE 9.2 SMLM on particles at interfaces. (a) experimental setup showing colloidal particles adsorbed at the interface of two liquids, on a coverslip. The geometry of the interface is probed by a dye that is dissolved in one of the phases and reversible adsorbs to the interface. This creates a signal akin to DNA-PAINT where temporal immobilization of the dye at the interface causes a fluorescence burst that can be localized. (b) Acquiring interfacial binding events over time enables the reconstruction of the geometry of the interface using principles borrowed from super-resolution localization microscopy. (c) this process can also be adopted to quantify interfacial properties for adsorbed elliptical particles. (Adapted with permission from Ref. [5].)

Subsequently, single particle contact angles, θ, were determined for both hydrophobic and hydrophilic spherical colloids at a water–octanol interface, with nanometer precision (Figure 9.2b). An effective line tension, τ, was inferred from the non-negligible dependence of θ on particle size as determined by these iPAINT experiments. Furthermore, the team demonstrated that interfacial deformations induced by adsorption of elliptical particles at water–decane interfaces could be visualized as well (Figure 9.2c).

Building upon the first SMLM study on particles at interfaces, a follow-up study investigated the microscopic origin of macroscopically observed emulsion phase inversions induced by a variation in particle size and aqueous phase pH [6]. The single particle contact angle, θ, of carboxyl polystyrene submicron particles (CPS) was found to decrease significantly upon an increase in aqueous phase pH and particle size, respectively (Figure 9.3). For CPS stabilized water-octanol emulsions, this allowed to switch emulsion type from water-in-oil to oil-in-water by adjustments in either particle size or pH.

These first super-resolution microscopy experiments on particles at interfaces [5, 6] clearly reveal the potential of these tools to help clarify hitherto unresolved physical origins of common observations related to the stability of particle-laden interfaces such as present in Pickering emulsions. An interesting future development would be to further develop the

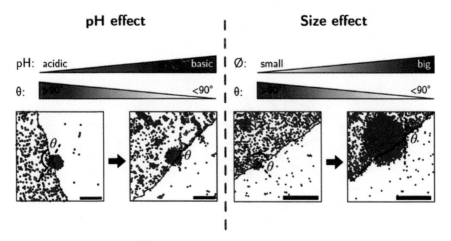

FIGURE 9.3 SMLM reveals that the single particle contact angle, θ, of carboxyl polystyrene submicron particles (CPS) decreases significantly upon an increase in aqueous phase pH and particle size. (Reprinted with permission from Ref. [6].)

methodology to enable a comprehensive and all-optical characterization of all relevant particle properties (size, shape [9], heterogeneities in surface chemistry [10], or surface roughness) and single-particle contact angles *in-situ* with parameterless interface detection [7]. Thus far, size and single-particle contact angle were determined within the exact same setup and sample in a particle-resolved manner, but particle shape, surface charge, and roughness cannot yet be resolved at the single-particle level. This would be particularly interesting to better understand stabilization with technical-grade or edible particles, which usually exhibit high degrees of variation in particle size, shape, and composition with major consequences for the local stabilization of (deformable) interfaces and hence overall stability and quality of, for example, low-fat dairy products [11] and membranes for separation technology [12].

9.3 MICROGELS

Microgels are a peculiar and exciting type of colloid [13]. They consist of a polymer gel network whose size is cut off on the micron or submicron scale, thus forming spherical microgel balls (see Figure 9.4). Commonly studied microgels have diameters ranging from the tens of nanometers to a micron or two. One can also make larger microgels, but then they gradually lose the characteristic colloidal properties, such as the large interfacial area and the dispersion homogeneity. Microgel colloids combine the polymer's macromolecular properties with a gel's topological features, all packed into a small colloidal unit. This combination of properties can be very beneficial for research and practical applications. One of the most

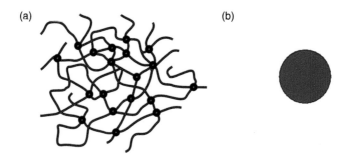

FIGURE 9.4   The picture in panel (a) shows a sketch of a polymer gel in which the linear homopolymer is crosslinked. The cross-linking sites are shown as small black circles. Bulk polymer gels are macroscopic, while the compacted microgels shown in (b) are typically 0.2–1 microns in diameter.

studied microgel colloids is made from the polymer poly(N-isopropylacrylamide) (PNIPAM) which one cross-links during the colloidal synthesis. PNIPAM is a particularly interesting polymer because it reacts to temperature changes. PNIPAM is water-soluble at room temperature but above $T_c = 32°C$ it is no longer soluble and the polymer precipitates. Microgels retain this property and show a volume phase transition at the same temperature. While at room temperature, the spongy particle is swollen and contains 80%–90% of water, almost all of the liquid is expelled above T_c as schematically indicated in Figure 9.4. Due to the fact that the particle size can be influenced or adjusted, microgel systems are often classified as intelligent materials because they can adapt to a changing environment.

Possible applications for microgels that are explored by researchers include their use as drug carriers with a controlled release mechanism, for tunable structurally colored opals, viscosity modifiers, sensors for trace amounts of molecules and contaminants, etc.

For all these possible applications, it is important to understand and characterize the structure and architecture of the microgels on the nanoscale, preferably in-situ and temperature-dependent. Often the microgels are also functionalized or hybridized by introducing additional nanoparticles such as SiO_2 or gold. The nanoparticles can themselves react to external influences or fulfil a specific purpose, as a colored (plasmonic) pigment, sensor or to reinforce the mechanical properties. For this case, traditional methods of characterization such as cryo-electron microscopy or light, X-ray, or neutron scattering are difficult to use and not specific. SMLM offers many advantages for studying microgels such as sufficient resolution (on the order of tens of nanometers), the ability for 3D imaging also of the internal structure of the microgel, and chemical specificity using fluorophore labeling. At the same time, microgels are easily labeled by fluorescent dyes because they are porous. Important open questions in relation to microgels concern the outwardly decreasing cross-linking density, the variations in properties within a population, and the packing behavior at high densities. For example, it was not clear for a long time whether microgels can interpenetrate under high (osmotic) pressure and if so, to what extent. In addition, the possibility that microgels could contract by themselves at high densities due to the increased concentration of co-ions (released by the microgels themselves) is still being discussed today.

With the help of SMLM, it has been possible to answer some of these open questions and address new ones in recent years. Conley et al.

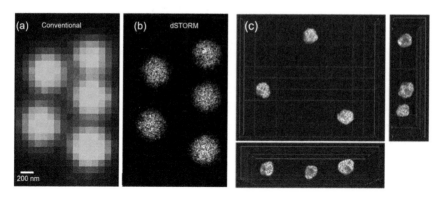

FIGURE 9.5   Microgels imaged with conventional wide-field microscopy (a) and with SMLM (b). The image in panel (c) shows a three-dimensional representation of heavily overpacked microgels where only a small quantity is labeled with fluorescent dye and visible in the image. (Adapted with permission from [14].)

demonstrated SMLM on microgels for the first time in 2016 [14] (Figure 9.5), shortly followed by Gelissen et al., who already investigated microgels decorated with gold nanoparticles [16]. While Gelissen compared SMLM with electron microscopy, Conley et al. compared their result with classical static and dynamic light scattering. They found good agreement concerning the properties that can be accessed by both methods, such as the core and shell size. Both groups used the dSTORM method, in which common fluorescent dyes are stochastically switched on and off with two lasers of different wavelengths. Subsequent studies by Conley et al. revealed packing behavior at ultra-high densities as shown in Figure 9.6 [15]. Using two-color SMLM, they showed that the outer layer of the colloidal microgel, a part that is not highly cross-linked, can indeed interpenetrate with nearest neighbors.

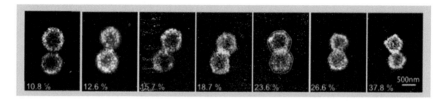

FIGURE 9.6   Two-color SMLM of densely packed microgels. The effective volume filling-fraction is eight times the mass filling-fraction listed in the image panels ranging from 86% to 300%. At densities above 15 wt%, the microgels deform and show facets. In addition, weak penetration can be observed, highlighted by the contour lines for 23.6wt%. (Adapted with permission from [15].)

The two-color labeling introduced in this first study also opened the door for future studies on microgels functionalized with different units, molecules, or nanoparticles, which can then be specifically resolved. All SMLM studies to date have shown that the microgels are nearly spherical and fairly homogeneous on the inside but have a core-shell structure with decreasing density toward the outside. Core and shell radius were uniformly distributed. Eisold and co-workers recently showed that DNA can be used to tune additional temperature sensitivity [17], which in turn allows switchable plasmon coupling and controlled uptake and release with the same system. SMLM has proven useful in the study of these smart materials and is likely to become a standard tool for the multi-color and 3D characterization of new colloidal materials.

REFERENCES

1. B. G. P. Van Ravensteijn, J. R. Magana Rodriguez, and I. K. Voets, "Manipulating matter with a snap of your fingers: a touch of Thanos in colloid science," *Superhero Sci. Technol.* **2**, 19–30 (2020).
2. M. Zanini, I. Lesov, E. Marini, C.-P. Hsu, C. Marschelke, A. Synytska, S. E. Anachkov, and L. Isa, "Detachment of rough colloids from liquid–liquid interfaces," *Langmuir* **34**, 4861–4873 (2018).
3. M. Zanini, A. Cingolani, C.-P. Hsu, M. Á. Fernández-Rodríguez, G. Soligno, A. Beltzung, S. Caimi, D. Mitrano, G. Storti, and L. Isa, "Mechanical phase inversion of Pickering emulsions via metastable wetting of rough colloids," *Soft Matter.* **15**, 7888–7900 (2019).
4. M. Rey, M. A. Fernandez-Rodriguez, M. Karg, L. Isa, and N. Vogel, "Poly-N-isopropylacrylamide nanogels and microgels at fluid interfaces," *Acc. Chem. Res.* **53**, 414–424 (2020).
5. A. Aloi, N. Vilanova, L. Isa, A. M. De Jong, and I. K. Voets, "Super-resolution microscopy on single particles at fluid interfaces reveals their wetting properties and interfacial deformations," *Nanoscale* **11**, (2019). 6654-6661
6. E. C. Giakoumatos, A. Aloi, and I. K. Voets, "Illuminating the impact of submicron particle size and surface chemistry on interfacial position and Pickering emulsion type," *Nano Lett.* **20**, 4837–4841 (2020).
7. D. L. H. van der Haven, R. P. Tas, P. van der Hoorn, R. van der Hofstad, and I. K. Voets, "Parameterless detection of liquid–liquid interfaces with submicron resolution in single-molecule localization microscopy," *J. Colloid Interface Sci.* **620**, 356–364 (2022).
8. A. Aloi, N. Vilanova, L. Albertazzi, and I. K. K. Voets, "iPAINT: a general approach tailored to image thetopology of interfaces with nanometer resolution †," *Nanoscale* **8**, 8712–8716 (2016).
9. A. Taylor, R. Verhoef, M. Beuwer, Y. Wang, and P. Zijlstra, "All-optical imaging of gold nanoparticle geometry using super-resolution microscopy," *J. Phys. Chem. C* **122**, 2336–2342 (2018).

10. R. M. Lubken, A. M. de Jong, and M. W. J. Prins, "How reactivity variability of biofunctionalized particles is determined by superpositional heterogeneities," *ACS Nano* **15**, 1331–1341 (2021).

11. C. C. Berton-Carabin and K. Schroën, "Pickering emulsions for food applications: background, trends, and challenges," *Ann. Rev. Food Sci. Technol.* **6**, 263–297 (2015).

12. M. F. Haase, K. J. Stebe, and D. Lee, "Continuous fabrication of hierarchical and asymmetric bijel microparticles, fibers, and membranes by solvent transfer-induced phase separation (STRIPS)," *Adv. Mater.* **27**, 7065–7071 (2015).

13. F. Scheffold, "Pathways and challenges towards a complete characterization of microgels," *Nat. Commun.* **11**, 4315 (2020).

14. G. M. Conley, S. Nöjd, M. Braibanti, P. Schurtenberger, and F. Scheffold, "Superresolution microscopy of the volume phase transition of pNIPAM microgels," *Colloids Surf. Physicochem. Eng. Asp.* **499**, 18–23 (2016).

15. G. M. Conley, P. Aebischer, S. Nöjd, P. Schurtenberger, and F. Scheffold, "Jamming and overpacking fuzzy microgels: deformation, interpenetration, and compression," *Sci. Adv.* **3**, e1700969 (2017).

16. A. P. H. Gelissen, A. Oppermann, T. Caumanns, P. Hebbeker, S. K. Turnhoff, R. Tiwari, S. Eisold, U. Simon, Y. Lu, J. Mayer, W. Richtering, A. Walther, and D. Wöll, "3D structures of responsive nanocompartmentalized microgels," *Nano Lett.* **16**, 7295–7301 (2016).

17. S. Eisold, L. H. Alvarez, K. Ran, R. Hengsbach, G. Fink, S. C. Benigno, J. Mayer, D. Wöll, and U. Simon, "DNA introduces an independent temperature responsiveness to thermosensitive microgels and enables switchable plasmon coupling as well as controlled uptake and release," *Nanoscale* **13**, 2875–2882 (2021).

CHAPTER 10

SRM Application to Supramolecular Structures

Deepika Gupta, Simanta Kalita, and Sarit S. Agasti

Jawaharlal Nehru Centre for Advanced Scientific Research (JNCASR), Bangalore, India

10.1 OVERVIEW

Supramolecular structures constructed by the non-covalent assembly of monomers have emerged over the last three decades as important nano-structured materials with promising applications in optoelectronics, light-harvesting, catalysis, and biology. In addition, the dynamic, transient, and adaptive nature of supramolecular nanoarchitectures showed great promise for use as a soft materials interface for creating biomimetic systems, catering to a variety of innovative and futuristic functions. Most notably, the function of these self-assembled supramolecular materials is highly dependent on their structure and dynamics. Therefore, to rationally design and develop supramolecular nanostructures with desired complexity and functional output, it is crucial to have an in-depth understanding of their structural characteristics and dynamic properties. Moreover, combining such knowledge with the mechanistic understanding of the self-assembly process can ultimately allow researchers to control molecular self-assembly at various length scales and hierarchical levels for designing next-generation functional supramolecular materials with advanced functionality. With the increasing push to understand supramolecular structures in greater detail, often, the use of conventionally employed

DOI: 10.1201/9781003220688-10

microscopy techniques such as transmission electron microscopy (TEM), atomic force microscopy (AFM), and scanning electron microscopy (SEM) proved challenging to shed light on many vital aspects of these nanostructures [1–4]. These include deconvoluting the organization complexity of multi-component systems, elucidating hierarchical structures and 3-dimensional (3D) networks, and probing real-time spatiotemporal growth, dynamics, and exchange kinetics of these supramolecular systems. In addition, lower sampling capability, invasive sample preparation procedures (e.g., drying or freezing), and the use of ionizing radiation that can damage or alter the sample itself, further limit the application of these imaging techniques for large-scale artifact-free investigation of the supramolecular structures under native conditions.

In this regard, super-resolution microscopy (SRM) proved to be a powerful optical technique that expanded the scientist's toolbox to complement and overcome the current constraints of conventional microscopy techniques. SRM is a sophisticated fluorescence-based imaging strategy that bridges the gap between high-resolution black-and-white imaging methods (such as TEM) and lower-resolution diffraction-limited multi-color imaging methods like Confocal Laser Scanning Microscopy (CLSM) [5–12]. The advent of super-resolution microscopy provides us with an unprecedented tool to gain detailed information about molecular assemblies with *in-situ* examining and visualization of supramolecular processes with greater precision and minimal invasiveness. Super-resolution microscopy techniques such as SMLM (STORM, PALM, and PAINT, see Chapter 2), STED (see Chapter 3), and SIM (see Chapter 4) have been materialized as a powerful tool to annotate the more detailed structural analysis and understand the dynamic characteristics of the supramolecular structures [3, 4]. The combination of nanoscopic resolution and multi-color monomers enabled researchers to study the complex 1D/2D/3D supramolecular assemblies of natural and synthetic supramolecular structures in greater detail. This chapter first discusses the influence of SRM on various single-component supramolecular systems as a means to elucidate the nanoscopic level structure and dynamics of the complex yet delicate artificial systems. Further, we illustrate multi-component supramolecular systems toward self-sorting and co-block polymerization that is rather difficult to decipher by other imaging techniques and paves the way toward relevant applications in energy materials and electronics.

10.2 SINGLE-COMPONENT SUPRAMOLECULAR ASSEMBLIES

10.2.1 Supramolecular Assemblies from Small Molecules

In 2014, for the first time, scientists employed an SMLM-based super-resolution imaging technique for the investigation of dynamics and monomer exchange mechanism in 1D supramolecular structures [13]. 1,3,5-benzenetricarboxamide (BTA) molecule assembles into 1D supramolecular nanofibers in aqueous solutions, in which incorporation of dye-labeled monomer (either BTA-Cy5 or BTA-Cy3) enabled STORM imaging of the nanofibers (Figure 10.1a). Through this method, scientists could visualize features of individual 1D BTA nanofibers down to ~25 nm. Further, the time-dependent monomer exchange among supramolecular fibers was monitored through a two-color STORM imaging experiment and stochastic simulation (Figure 10.1a). For this purpose, two individually labeled single-color fibers (labeled with BTA-Cy5 and BTA-Cy3) were mixed, and the exchange of monomers was monitored by imaging the migration of Cy5-labeled BTA monomers to a Cy3-labeled fiber and vice versa (Figure 10.1b). Notably, here the ability to probe the monomer distribution at different time points using STORM imaging allowed researchers to quantitatively address the mechanism of BTA monomer exchange between fibers. These imaging experiments unambiguously depicted random monomer exchange along the fibers and ruled out the possibility that monomer exchange takes place only from ends or via breakage and recombination mechanism. In addition, the consequences of various factors such as chirality and functional groups in BTA-based supramolecular polymers were also investigated by the two-color STORM method [14, 15]. Other than BTAs, locating and tracking specific monomers in a mixture of distinctly labeled ureidopyrimidone (Upy)-based polymers have also been studied using STORM technique [16]. It was concluded that the structure and exchange dynamics of self-assembled nanofibers depend heavily on the design of UPy monomers and can also be controlled by mixing different monomer variants. The above examples manifest that SRM with STORM technique can efficiently reveal the structure-dynamic relationship and investigate the mechanism and kinetics of monomer exchange in small molecules supramolecular polymers. Other than STORM, the dynamic synthetic supramolecular nanostructures in organic solvents can also be investigated by PALM imaging [17].

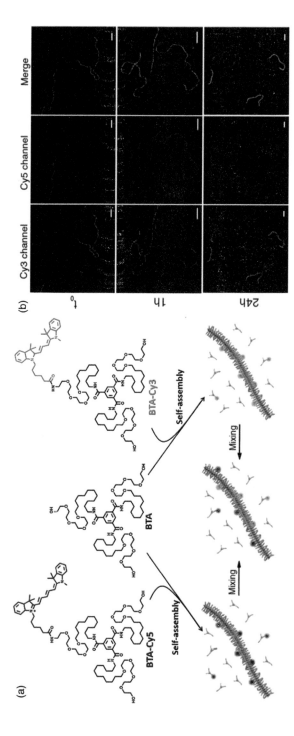

FIGURE 10.1 (a) Molecular structure of monomer (BTA) and fluorescence labels (BTA-Cy3, and BTA-Cy5) and schematic representation of monomer exchange on mixing aqueous solutions of differently labeled supramolecular polymers. (b) STORM images obtained after mixing red (Cy5) and green (Cy3) supramolecular polymers as a function of time. Scale bars: 1 μm. (Reproduced with permission from Ref. [13].)

10.2.2 Supramolecular Assemblies from Peptides and Proteins

Self-assembled structures of peptide amphiphiles give exciting opportunities to be used as artificial functional materials. However, nanoscale-level information on structure and dynamics is necessary for optimizing its function and performance. Similar to small molecules, the molecular exchange mechanism among self-assembled peptide (PA) nanofibers was probed by two-color STORM imaging [18]. It involved time-dependent investigation of two different fluorescently labeled fibers, PA-Cy3 and PA-Cy5, mixed with unlabeled monomer PA. Remarkably, it was observed that various intermolecular interactions operating at the level of an individual peptide nanofiber affect the exchange rates within the same fiber. The STORM-based time-lapse imaging experiment concluded the coexistence of structural diversity from fully dynamic to kinetically inactive areas in the supramolecular architecture. STORM imaging is also used to investigate "living" fibril formation in the histidine-containing triblock protein polymer [19]. In their molecular design, the monomeric unit is constituted by a silk-inspired central block of GAGAGAGH repeats and hydrophilic collagen-based domains at the periphery. These monomers are folded and stacked to form nanofibrils, and further, their exchange dynamics are probed by mixing differently labeled (Alexa Fluor 488 and Alexa Fluor 647) monomers. Probing individual nanofibers as a function of time depicted negligible exchange even after 3 days (Figure 10.2a). Moreover, they also studied the "living" nature of these stable protein nanofibers and their directional growth by directly mixing fresh dye-labeled monomeric units into mature nanofibers tethered to another dye molecule. Surprisingly and interestingly, out of two reactive fiber ends, fiber growth is encouraged from only one end resulting in unidirectional nanofibers (Figure 10.2b). Hence, diblock protein nanofibers with controlled length were achieved on sequential and controlled addition of distinctly fluorescent labeled protein polymers. STORM technique was further employed to study the self-assembly of I_3K peptide fibers in the bulk phase [20]. The peptide amphiphile self-assembles resulting in helical fibrils (diameter 5–10 nm) with significant helical twisting. Two-color STORM technique was utilized to gain quantitative insight into the monomer-exchange dynamics. Such supramolecular structures were stable and did not show significant monomer exchange. Remarkably, the dynamic morphological changes of the enzyme-instructed supramolecular assemblies inside cancer cells can also be investigated using

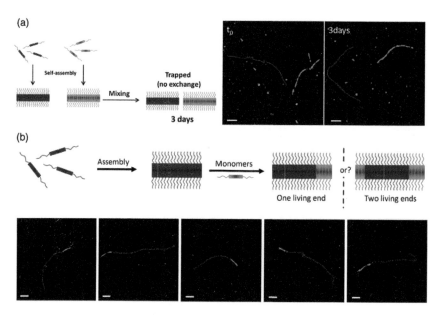

FIGURE 10.2 (a) Monomer exchange in fibrils composed of a *de novo*-designed recombinant triblock protein polymer. STORM image after 3 days indicates that no exchange takes place between fibers. (b) STORM images of fibrils obtained from the sequential monomer addition showed a number of red–green diblock structures, indicating the existence of a living end. Scale bars: 1 μm (a) and 1 μm (b). (Reproduced with permission from Ref. [19].)

dSTORM with a resolution below 50 nm [21]. This method allowed the characterization of the spatial distribution and morphology transformation of supramolecular assemblies in live cells, which endure drastically different pathways between cell lines, leading to cell-selective biological activities. Other than the STORM technique, the hydrogels formed from minimalistic peptide amphiphiles (such as Fmoc-FF and FF) can also be imaged by PAINT technique [22, 23]. Here, dye labeled monomer (e.g., Cy5-FF) acted as PAINT probe and displayed the ability to reversibly bind to the self-assembled nanostructures, enabling precise localization of the binding events (Figure 10.3a). Through this strategy, the 2D and 3D fiber networked hydrogel can be visualized with improved spatial resolution, enabling nanoscopic quantification of fiber diameter, distribution, and mesh size of the network (Figure 10.3b and c).

In the above examples, dye-labeled monomer design is pivotal for the nanoscale imaging of supramolecular structures. Alternatively, a general labeling approach, circumventing the specific covalent modifications of

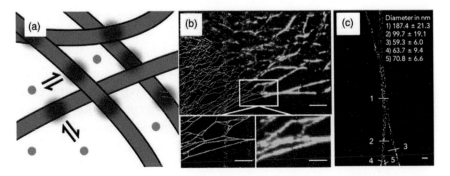

FIGURE 10.3 (a) Strategy for PAINT imaging with labeled monomer. (b) Time-dependent supramolecular nanostructures of FF doped with dye-peptides. (b) PAINT image of Fmoc-FF hydrogel with Cy5-FF as PAINT probe. (c) Zoomed image of nanofiber depicting high resolution of PAINT technique. Scale bars: 5 μm (b, top image), 2.5 μm (b, bottom image), and 200 nm (c). (Reproduced with permission from Ref. [23].)

the monomer, has also been developed for SRM imaging of supramolecular structures. In this regard, STED-based SRM imaging was employed for investigating a supramolecular peptide nanostructure based on non-specific electrostatic interactions between the fluorescent label and the self-assembled structure [24]. The peptide nanostructure having cationic residues and fluorophore with anionic motifs (Alexa-488) electrostatically binds and enables the direct visualization of hierarchical supramolecular fibers via STED imaging. In addition, the researchers showed that similar STED imaging could also be employed to investigate a dynamic disassembly process, where enzymatic degradation of peptide nanofibers was monitored in real-time and *in-situ*. In another approach, encapsulation of Nile Red into the hydrophobic environment of the supramolecular polymer was explored for SIM-based super-resolution imaging of supramolecular assemblies [25]. Through this approach, the researchers investigated the formation of supramolecular polymers from intramolecular charge transfer (CT) foldamer. The CT foldamer (PNF) structure consists of pyrene as a donor component and flexibility attached naphthalene diimide (NDI) as an acceptor component, which exists in foldamer confirmation via strong intramolecular CT interactions. The folded amphiphiles self-assemble into one-dimensional (1D) supramolecular polymers with stacked bilayers, allowing an efficient encapsulation of Nile Red dye and visualization under SIM. Further, researchers visualized the kinetically controlled growth process using the redox-responsive dormant states that yielded monodisperse CT supramolecular polymers with a high degree of polymerization.

In most cases discussed above, a fluorescent label with the desired photophysical property was usually tethered to the self-assembling building blocks or with the supramolecular structures. However, as supramolecular assemblies are sensitive to every minute change in composition and environmental conditions, in certain cases, the effect of fluorescent dye or imaging conditions (e.g., specialized STORM imaging buffer) on the self-assembly process may also become significant. In this regard, core substituted naphthalene diamide (cNDI)-peptide amphiphile conjugates provide an interesting self-assembling monomeric unit with an inherent fluorescent property. Supramolecular assemblies from these cNDI amphiphile exhibit fluorescent assemblies without the aid of external fluorophores [26]. Further, multi-color super-resolution investigation based on SIM microscopy was possible based on differently substituted cNDI, emitting at different wavelengths. More recently, researchers exploited the photophysical properties of the stacked monomers for nanoscale analysis of the structure, heterogeneity, and kinetics of the supramolecular assemblies from the benzyl-naphthalimide dyes [27].

10.3 SELF-SORTED MULTI-COMPONENT SYSTEMS

A two-component system unveils the exciting possibilities for generating double-networked architectures by mixing two orthogonal monomers. Such monomers self-sort their liked ones, resulting in pure individual assemblies. However, it is difficult to elucidate the self-sorting phenomenon by traditional ensemble spectroscopic techniques as differentiation of the two different building blocks from bulk measurement is challenging. The super-resolution fluorescence microscopy provides the much-needed multi-color visualizing capability at nanoscopic resolution for investigating self-sorted supramolecular assemblies. In 2016, the first example of *in-situ* real-time visualization of self-sorted fibers entangled in 2D and 3D was reported via STED imaging (Figure 10.4a) [28]. Self-sorted hydrogels consisting of two distinct building blocks, peptide-based gelator and lipid-like hydrogelator, were tagged with fluorescent probes capable of selectively staining the individual fibers, facilitating their multi-color visualization under super-resolution microscope (STED). STED imaging depicted an extremely sharp image of fibers with ~80 nm in diameter, spatially arranged orthogonal to one another in the cross-sectional view (Figure 10.4b). In this multi-component system, fiber formation

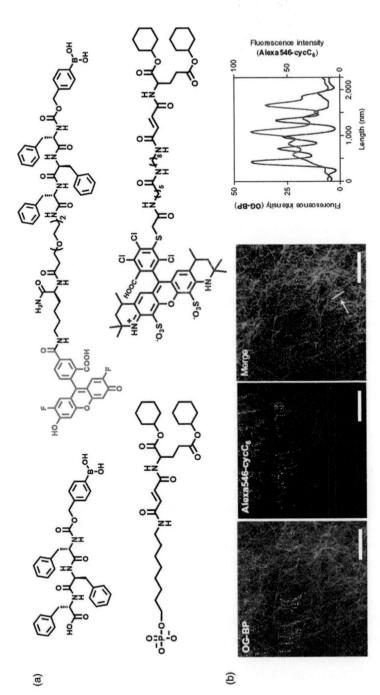

FIGURE 10.4   (a) Molecular structures of the peptide-based and lipid-based hydrogelators and their corresponding selective fluorescent stains. (b) STED images were recorded after mixing both hydrogelators and fluorescent probes. The merged image and plot of fluorescent intensity clearly indicate the orthogonal self-assembly. Scale bars: 5 µm. (Reproduced with permission from Ref. [1].)

occurs through a cooperative mechanism, and fibers with different building blocks have different formation rates. In addition, because of their orthogonal assembly, the respective physicochemical properties of the two fibers remain intact even after self-sorting. The SIM technique was also employed to directly visualize self-sorted amyloid-inspired minimalistic peptide amphiphiles, C_{10}-VFFAKK [29]. The enantiomeric hexapeptides are revealed to exhibit chirality-driven self-sorted nanofibers that further showed enantioselective enzymatic degradation for *L*-peptide fibers.

10.4 SUPRAMOLECULAR BLOCK ASSEMBLIES

One of the most interesting supramolecular structures at the nanoscale is block co-polymers, consisting of distinct blocks originating from different monomers. Supramolecular block co-polymers are an attractive class of light-harvesting organic heterostructures that are becoming attractive materials for many exciting applications. The information about the structural properties of such co-polymers is essential to tune and engineer these systems to achieve desired properties. For that super-resolution, microscopy has established itself as a very vital tool for investigating structural organization. Intrinsic "color-coded" emission from each monomer, enabling multi-color super-resolution imaging, promises to provide exciting structural insight at the nanoscale level. Super resolution-SIM microscopy has been recently exploited to understand block-copolymer formation from inherently emissive cNDI monomers [30]. Achieving narcissistically self-sorted homopolymers, random, and block co-polymer sequences from the same monomer pair, and their dynamic reconfiguration is an extremely challenging task. Through appropriate usage of kinetic and thermodynamic pathway complexity of molecular self-assembly, researchers targeted this challenge and accomplished an unprecedented sequence control in the supramolecular co-polymerization of carbonate cholesterol appended two cNDI monomers to access self-sorted, random, and block supramolecular polymers (Figure 10.5a). Symmetrically disubstituted NDI core with ethanethiol (1) and ethoxy (2) groups gave rise to distinctly emissive monomers (570–750 nm for ethanethiol and 450–570 nm for ethoxy), which allowed orthogonal probing of its self-assembly characteristics and characterization of the multi-component structures by visualizing under super-resolved SIM imaging (Figure 10.5b). In a subsequent study, researchers reported a co-operative supramolecular block co-polymerization of fluorescent cNDI donor and acceptor monomers in solution under thermodynamic

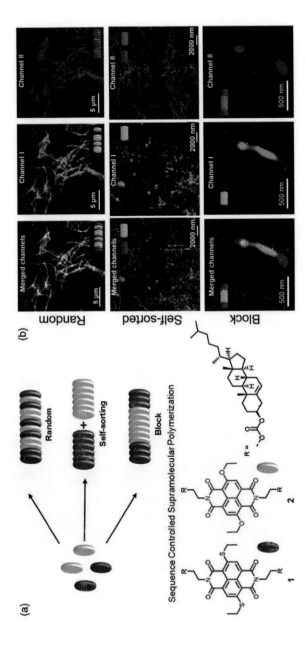

FIGURE 10.5 (a) Schematic representation of two-component supramolecular polymerization into self-sorted, random, and block supramolecular polymers. Molecular structures of cNDI monomers that are used for this two-component supramolecular polymerization. (b) SIM images of self-sorted, random, and block supramolecular polymers. Scale bars are given in the images. (Reproduced with permission from Ref. [31].)

control for the synthesis of axial organic heterostructures with the efficient light-harvesting property [31]. Ethoxy and pentane thiol cNDI derivatives have been used as green and red fluorescence emitters, and to facilitate a hydrogen-bonded co-operative supramolecular polymerization the cNDI derivatives were attached to chiral trans-1,2-bis(amido)-cyclohexane motif. The resultant bischromophoric derivatives (SS-dithiol and SS-diOEt) with minimal structural mismatch were heated in a suitable solvent to 363 K and cooled under thermodynamic control at a rate of 1 K/min, which exhibited heterogeneous nucleation of SS-diOEt on SS-dithiol nuclei (Figure 10.6a). The distinct green and red fluorescence of the SS-diOEt and SS-dithiol stacked monomers enabled super-resolved SIM visualization of the co-assembled solution, revealing blocky supramolecular polymer structure with alternating green and red segments of stacked SS-diOEt and SS-dithiol monomers (Figure 10.6b). In addition, the high-resolution fluorescence imaging allowed researchers to reliably analyze the distribution of individual blocks, which showed a narrow polydispersity with a PDI of 1.1 for the individual block segments.

Most supramolecular multi-component systems are limited to two functional monomers; however, the advances in nanometer-level multiplexed imaging promise the study of more complex tricomponent systems.

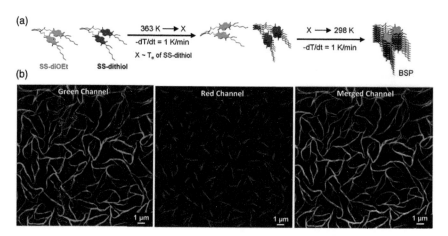

FIGURE 10.6 (a) Schematics depicting the process of formation of block co-polymers. The monomers were annealed and cooled under thermodynamic control until the elongation temperature (T_e) of SSdithiol triggered the heterogeneous nucleation of **SS-diOEt** in **SS-dithiol** seeds. (b) SIM microscopy image depicts the formation of supramolecular block co-polymers. Scale bars are given in the images. (Reproduced with permission from Ref. [30].)

Researchers recently established a tricomponent system of different cNDI coupled with diphenylalanine peptides having three distinct emissions [32]. Two monomers with derivatives of ethoxy (1) and ethanethiol (2) have been discussed already; the third emissive monomer has a core substitution of isopropyl amine (3) (Figure 10.7a). Using these monomers, supramolecular block co-polymers with tunable microstructure were achieved via kinetically controlled, sequential seed-induced living supramolecular polymerization, which was probed in detail through spectroscopic studies (Figure 10.7b). Further, these monomers' distinct blue, green, and red fluorescent nature aided the SIM visualization of segmented topology from these multi-component axial heterostructures (Figure 10.7c). Interestingly, by varying the seeds and order of monomer addition, different sequences of tricomponent supramolecular block co-polymers were achieved.

To design diverse higher-order complex nanostructures, understanding and achieving control over the molecular organization at a higher hierarchical level of self-assembly is essential. In this regard, the secondary nucleation events in synthetic supramolecular systems open up exciting possibilities with unprecedented nanostructure control [33]. Using SIM, researchers recently investigated molecular chirality control on the primary and secondary nucleation events in seed-induced supramolecular polymerization. This study performed homochiral and heterochiral seeding experiments with enantiomeric chiral binaphthyl appended bis-NDI with different substituents. In the case of homochiral seeding, the primary nucleation–elongation process dominates, rendering the block co-polymer formation. On the other hand, the secondary nucleation process dominates heterochiral seeding, resulting in self-sorted supramolecular polymers. Such strong homochiral recognition abilities to result in block co-polymer and surface-catalyzed self-sorted polymers mediated by secondary nucleation were visualized by SIM imaging.

Unfortunately, most of the SRM techniques require dye functionalization to the monomer or synthesis of specific monomers with suitable photophysical properties. In this regard, super-resolution imaging with the PAINT technique involving transient non-covalent staining avoiding monomer modification, enables us to study self-assembled nanostructures with higher spatial resolution. Inspired by PAINT and PALM, researchers recently introduced a combinatorial extension of these techniques (coined iPAINT (interface point accumulation for imaging in nanoscale topography)) for 3D, sub-diffraction imaging of solid/liquid, liquid/liquid, and liquid/air interfaces [34, 35]. In recent years, researchers explored the use

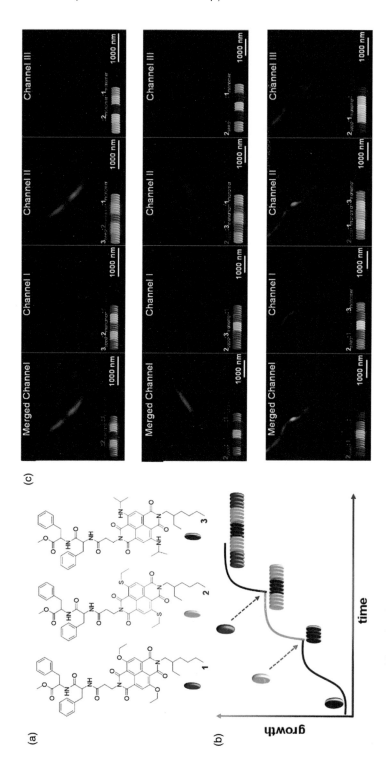

FIGURE 10.7 (a) Molecular structure of diphenylalanine anchored naphthalene diimide core with ethoxy (**1**), ethanethiol (**2**) and isopropylamine (**3**) substitution. (b) Kinetic strategy for the formation of the tricomponent supramolecular block co-polymer. (c) SIM images of three-component supramolecular block co-polymers with tunable microstructure. (Reproduced with permission from Ref. [32].)

of iPAINT to investigate supramolecular polymer structure in nonpolar organic media (Figure 10.8a) [35]. In addition, the researchers cleverly used the quasi-permanent adsorption of the dye to the polymer structure to identify block-like arrangements within supramolecular fibers. In this case, block supramolecular fibers were obtained upon mixing homopolymers that were stained previously with different caged fluorophores. An example takes into account of the formation of the block copolymer poly[(S-1)-co-(S-2)] made of homopolymer poly(S-1) and homopolymer poly(S-2). Poly(S-1), pre-stained with Cage-635 dye forms the block copolymer poly[(S-1)-co-(S-2)] when it is mixed with Poly(S-2) pre-stained with Cage552 dye at 40°C in 1:1 ratio. The block copolymer formed can be imaged using i-PAINT, as the dyes Cage635 and Cage552 are already present on their surface (Figure 10.8b).

In recent times, living crystallization-driven self-assembly (CDSA), a seeded-growth method for block co-polymers (BCPs) with a crystallizable core-forming block, has emerged as a highly promising and versatile route to core–shell nanoparticles with high levels of precision and control [36–38]. Recently, researchers used SRM techniques to investigate the

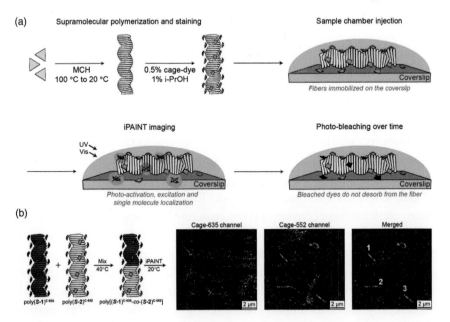

FIGURE 10.8 (a) Schematic representation of iPAINT workflow. (b) Schematic representation of the iPAINT image of the formed block co-polymer poly[(S-1)-co-(S-2)]. (Reproduced from Ref. [35] (open access).)

kinetics of this seeded growth under a variety of conditions for poly-(ferrocenyldimethylsilane)-b-(polydimethylsiloxane) (PFS-b-PDMS) BCP, a model living CDSA system [38]. For this purpose, fluorescently labeled BCP unimer (with STAR635 dye for single-color STED and CAGE500 for dual-color SMLM) was added to a short fluorescent seed micelles solution that was prepared from BCPs and labeled with the appropriate dye (STAR635 for single-color STED and CAGE635 for dual-color SMLM). Aliquots were taken at different time points and were visualized by STED and SMLM to gain quantitative and detailed information on the living CDSA growth process. SMLM and STED techniques were useful for quantitatively analyzing such blocks to measure the size and get information about the extent of control being achieved. Also, SMLM aids in in-situ visualization of such block co-polymerization processes, which is very useful in understanding the kinetics and growth bias toward a particular direction. An important point to keep in mind is that dyes used for the imaging can have an impact on the growth kinetics of the block co-polymer, though it can be minimal. Howsoever, super-resolution microscopy has become a boon to scientists in the supramolecular chemistry arena, and it is consolidating its position in recent times at an ever-increasing speed.

10.5 CONCLUSION AND FUTURE PERSPECTIVE

SRM has already imprinted its footprint in the biological realm. Bypassing the diffraction barrier of light for optical microscopy, SRM enabled researchers to have detailed insight into intracellular structures and various dynamic processes in the living system in an unprecedented way. By following up on this advancement, the supramolecular chemistry community is also slowly and steadily adopting the approach of SRM as an alternative tool for structural and dynamics characterization of supramolecular assemblies. In many ways, the SRM complements and overcomes many constraints of conventional microscopy techniques like TEM, AFM, and SEM. More specifically, SRM significantly contributed to the detailed understanding of the structures of multi-component systems and probing the dynamics of supramolecular assemblies. Given the delicate nature of the non-covalent bonding and supramolecular assemblies, it is necessary to carefully select the appropriate labeling technique and SRM method that has minimal influence on the assembly. This might render certain newly developed dyes with highly optimized photophysical properties for SRM to be out of favor for imaging supramolecular structures. Therefore, there is room for exploring new labeling strategies and developing minimally perturbing probe structures for achieving better spatiotemporal resolution from SRM-based investigation of supramolecular structures. The intrinsic fluorescence from the monomers provides an alluring solution to investigate native supramolecular structures without any additional labeling perturbation. However, these monomers may not be optimized for providing high imaging resolution, and such strategies are usually not generalizable to other supramolecular structures. In this context, it should be noted that with the appropriate use of fluorescent probes and SRM techniques, it is now possible to achieve close to the molecular level of resolution [39, 40]. With more and more interest in using SRM for supramolecular chemistry, hopefully, these methods will soon make an impact in this area. Another important point that needs special attention is having real-time dynamic information on the supramolecular systems. Most of the dynamic processes happen in the bulk solution, whereas super-resolution microscopy operates close to the coverslip. Researchers take out aliquots of the mixture and deposit them on the coverslip to image; this leads to a decreased temporal resolution. Also, direct solution state measurement is prevented by the dyes, which offer a very high background. So, strategies that can alleviate these challenges will be of great importance. Overall, in the materials science regime, SRM is still in the nascent stage, which is yet to be explored and exploited more. With the increasing interest of researchers in using SRM for supramolecular chemistry, it can be expected that SRM will become a new regular technique for supramolecular chemistry research.

REFERENCES

1. R. Kubota, W. Tanaka, and I. Hamachi, "Microscopic imaging techniques for molecular assemblies: electron, atomic force, and confocal microscopies," *Chem. Rev.* **121**, 14281–14347 (2021).
2. A. Rizvi, J. T. Mulvey, B. P. Carpenter, R. Talosig, and J. P. Patterson, "A close look at molecular self-assembly with the transmission electron microscope," *Chem. Rev.* **121**, 14232–14280 (2021).
3. S. Dhiman, T. Andrian, B. S. Gonzalez, M. M. E. Tholen, Y. Wang, and L. Albertazzi, "Can super-resolution microscopy become a standard characterization technique for materials chemistry?," *Chem. Sci.* **13**, 2152–2166 (2022).
4. S. Pujals, N. Feiner-Gracia, P. Delcanale, I. Voets, and L. Albertazzi, "Super-resolution microscopy as a powerful tool to study complex synthetic materials," *Nat. Rev. Chem.* **3**, 68–84 (2019).

5. S. W. Hell, "Far-field optical nanoscopy," *Science* **316**, 1153–1158 (2007).

6. K. Xu, H. P. Babcock, and X. Zhuang, "Dual-objective STORM reveals three-dimensional filament organization in the actin cytoskeleton," *Nat. Methods* **9**, 185–188 (2012).

7. M. Lelek, M. T. Gyparaki, G. Beliu, F. Schueder, J. Griffié, S. Manley, R. Jungmann, M. Sauer, M. Lakadamyali, and C. Zimmer, "Single-molecule localization microscopy," *Nat. Rev. Methods Primers* **1**, 1–27 (2021).

8. J. Schnitzbauer, M. T. Strauss, T. Schlichthaerle, F. Schueder, and R. Jungmann, "Super-resolution microscopy with DNA-PAINT," *Nat. Protoc.* **12**, 1198–1228 (2017).

9. R. Jungmann, M. S. Avendaño, J. B. Woehrstein, M. Dai, W. M. Shih, and P. Yin, "Multiplexed 3D cellular super-resolution imaging with DNA-PAINT and Exchange-PAINT," *Nat. Methods* **11**, 313–318 (2014).

10. R. Sasmal, N. D. Saha, F. Schueder, D. Joshi, V. Sheeba, R. Jungmann, and S. S. Agasti, "Dynamic host–guest interaction enables autonomous single molecule blinking and super-resolution imaging," *Chem. Commun.* **55**, 14430–14433 (2019).

11. T. Andrian, P. Delcanale, S. Pujals, and L. Albertazzi, "Correlating super-resolution microscopy and transmission electron microscopy reveals multi-parametric heterogeneity in nanoparticles," *Nano Lett.* **21**, 5360–5368 (2021).

12. J. Yan, L.-X. Zhao, C. Li, Z. Hu, G.-F. Zhang, Z.-Q. Chen, T. Chen, Z.-L. Huang, J. Zhu, and M.-Q. Zhu, "Optical nanoimaging for block copolymer self-assembly," *J. Am. Chem. Soc.* **137**, 2436–2439 (2015).

13. L. Albertazzi, D. van der Zwaag, C. M. A. Leenders, R. Fitzner, R. W. van der Hofstad, and E. W. Meijer, "Probing exchange pathways in one-dimensional aggregates with super-resolution microscopy," *Science* **344**, 491–495 (2014).

14. M. B. Baker, L. Albertazzi, I. K. Voets, C. M. A. Leenders, A. R. A. Palmans, G. M. Pavan, and E. W. Meijer, "Consequences of chirality on the dynamics of a water-soluble supramolecular polymer," *Nat. Commun.* **6**, 6234 (2015).

15. M. B. Baker, R. P. J. Gosens, L. Albertazzi, N. M. Matsumoto, A. R. A. Palmans, and E. W. Meijer, "Exposing differences in monomer exchange rates of multicomponent supramolecular polymers in water," *ChemBiochem* **17**, 207–213 (2016).

16. S. I. S. Hendrikse, S. P. W. Wijnands, R. P. M. Lafleur, M. J. Pouderoijen, H. M. Janssen, P. Y. W. Dankers, and E. W. Meijer, "Controlling and tuning the dynamic nature of supramolecular polymers in aqueous solutions," *Chem. Commun.* **53**, 2279–2282 (2017).

17. A. Aloi, A. Vargas Jentzsch, N. Vilanova, L. Albertazzi, E. W. Meijer, and I. K. Voets, "Imaging nanostructures by single-molecule localization microscopy in organic solvents," *J. Am. Chem. Soc.* **138**, 2953–2956 (2016).

18. R. M. P. da Silva, D. van der Zwaag, L. Albertazzi, S. S. Lee, E. W. Meijer, and S. I. Stupp, "Super-resolution microscopy reveals structural diversity in molecular exchange among peptide amphiphile nanofibres," *Nat. Commun.* **7**, 11561 (2016).

19. L. H. Beun, L. Albertazzi, D. van der Zwaag, R. de Vries, and M. A. Cohen Stuart, "Unidirectional living growth of self-assembled protein nanofibrils revealed by super-resolution microscopy," *ACS Nano* **10**, 4973–4980 (2016).

20. H. Cox, P. Georgiades, H. Xu, T. A. Waigh, and J. R. Lu, "Self-assembly of mesoscopic peptide surfactant fibrils investigated by STORM super-resolution fluorescence microscopy," *Biomacromolecules* (2017).

21. Q. Yao, C. Wang, M. Fu, L. Dai, J. Li, and Y. Gao, "Dynamic detection of active enzyme instructed supramolecular assemblies *in situ* via super-resolution microscopy," *ACS Nano* **14**, 4882–4889 (2020).

22. S. Pujals, K. Tao, A. Terradellas, E. Gazit, and L. Albertazzi, "Studying structure and dynamics of self-assembled peptide nanostructures using fluorescence and super resolution microscopy," *Chem. Commun.* **53**, 7294–7297 (2017).

23. E. Fuentes, K. Boháčová, A. M. Fuentes-Caparrós, R. Schweins, E. R. Draper, D. J. Adams, S. Pujals, and L. Albertazzi, "PAINT-ing fluorenylmethoxy-carbonyl (Fmoc)-diphenylalanine hydrogels," *Chem. Eur. J.* **26**, 9869–9873 (2020).

24. M. Kumar, J. Son, R. H. Huang, D. Sementa, M. Lee, S. O'Brien, and R. V. Ulijn, "In situ, noncovalent labeling and stimulated emission depletion-based super-resolution imaging of supramolecular peptide nanostructures," *ACS Nano* **14**, 15056–15063 (2020).

25. K. Jalani, A. D. Das, R. Sasmal, S. S. Agasti, and S. J. George, "Transient dormant monomer states for supramolecular polymers with low dispersity," *Nat. Commun.* **11**, 3967 (2020).

26. A. Sarkar, J. C. Kölsch, C. M. Berač, A. Venugopal, R. Sasmal, R. Otter, P. Besenius, and S. J. George, "Impact of NDI-core substitution on the pH-responsive nature of peptide-tethered luminescent supramolecular polymers," *ChemistryOpen* **9**, 346–350 (2020).

27. Q. Qiao, W. Liu, Y. Zhang, J. Chen, G. Wang, Y. Tao, L. Miao, W. Jiang, K. An, and Z. Xu, "In situ real-time nanoscale resolution of structural evolution and dynamics of fluorescent self-assemblies by super-resolution imaging," *Angew. Chem. Int. Ed.* **61**, e202208678 (2022).

28. S. Onogi, H. Shigemitsu, T. Yoshii, T. Tanida, M. Ikeda, R. Kubota, and I. Hamachi, "In situ real-time imaging of self-sorted supramolecular nanofibres," *Nat. Chem.* **8**, 743–752 (2016).

29. D. Gupta, R. Sasmal, A. Singh, J. P. Joseph, C. Miglani, S. S. Agasti, and A. Pal, "Enzyme-responsive chiral self-sorting in amyloid-inspired minimalistic peptide amphiphiles," *Nanoscale* **12**, 18692–18700 (2020).

30. A. Sarkar, T. Behera, R. Sasmal, R. Capelli, C. Empereur-Mot, J. Mahato, S. S. Agasti, G. M. Pavan, A. Chowdhury, and S. J. George, "Cooperative supramolecular block copolymerization for the synthesis of functional axial organic heterostructures," *J. Am. Chem. Soc.* **142**, 11528–11539 (2020).

31. A. Sarkar, R. Sasmal, C. Empereur-Mot, D. Bochicchio, S. V. K. Kompella, K. Sharma, S. Dhiman, B. Sundaram, S. S. Agasti, G. M. Pavan, and S. J. George, "Self-sorted, random, and block supramolecular copolymers via sequence controlled, multicomponent self-assembly," *J. Am. Chem. Soc.* **142**, 7606–7617 (2020).

32. A. Sarkar, R. Sasmal, A. Das, A. Venugopal, S. S. Agasti, and S. J. George, "Tricomponent supramolecular multiblock copolymers with tunable composition via sequential seeded growth," *Angew. Chem. Int. Ed.* **60**, 18209–18216 (2021).

33. S. Sarkar, A. Sarkar, A. Som, S. S. Agasti, and S. J. George, "Stereoselective primary and secondary nucleation events in multicomponent seeded supramolecular polymerization," *J. Am. Chem. Soc.* **143**, 11777–11787 (2021).

34. A. Aloi, N. Vilanova, L. Albertazzi, and I. K. Voets, "iPAINT: a general approach tailored to image the topology of interfaces with nanometer resolution," *Nanoscale* **8**, 8712–8716 (2016).

35. B. Adelizzi, A. Aloi, N. J. Van Zee, A. R. A. Palmans, E. W. Meijer, and I. K. Voets, "Painting supramolecular polymers in organic solvents by super-resolution microscopy," *ACS Nano* **12**, 4431–4439 (2018).

36. J. A. Massey, K. Temple, L. Cao, Y. Rharbi, J. Raez, M. A. Winnik, and I. Manners, "Self-assembly of organometallic block copolymers: the role of crystallinity of the core-forming polyferrocene block in the micellar morphologies formed by poly(ferrocenylsilane-b-dimethylsiloxane) in n-alkane solvents," *J. Am. Chem. Soc.* **122**, 11577–11584 (2000).

37. G. S. Kaminski Schierle, S. van de Linde, M. Erdelyi, E. K. Esbjörner, T. Klein, E. Rees, C. W. Bertoncini, C. M. Dobson, M. Sauer, and C. F. Kaminski, "In situ measurements of the formation and morphology of intracellular β-amyloid fibrils by super-resolution fluorescence imaging," *J. Am. Chem. Soc.* **133**, 12902–12905 (2011).

38. C. E. Boott, E. M. Leitao, D. W. Hayward, R. F. Laine, P. Mahou, G. Guerin, M. A. Winnik, R. M. Richardson, C. F. Kaminski, G. R. Whittell, and I. Manners, "Probing the growth kinetics for the formation of uniform 1D block copolymer nanoparticles by living crystallization-driven self-assembly," *ACS Nano* **12**, 8920–8933 (2018).

39. M. Weber, H. von der Emde, M. Leutenegger, P. Gunkel, S. Sambandan, T. A. Khan, J. Keller-Findeisen, V. C. Cordes, and S. W. Hell, "MINSTED nanoscopy enters the Ångström localization range," *Nat. Biotechnol.* **41**, 569–576 (2023).

40. M. Dai, R. Jungmann, and P. Yin, "Optical imaging of individual biomolecules in densely packed clusters," *Nat. Nanotechnol.* **11**, 798–807 (2016).

CHAPTER 11

Super-Resolution Microscopy Application to Nanomedicine

Sílvia Pujals

Institute for Advanced Chemistry of Catalonia (IQAC-CSIC), Barcelona, Spain

In view of the challenges that nanomedicine is facing, super resolution microscopy (SRM) could be a technique fulfilling some of the gaps and needs of nanomedicine, easing the standardization of nanomaterials characterization, and providing protocols for establishing nano-bio interactions. Thanks to reaching the nanoscale, super resolution microscopy is ideal to study nanomaterials, from characterizing them in vitro to learning details about their performance in the biological context (nano-bio interaction).

SRM is ideally positioned to elucidate key nanomaterial properties like size and shape, size dispersity and aggregation, zeta potential, composition, density, drug loading and release, targeting, and labeling, with the possibility of giving such information at the single particle and single molecule level. The amount of work done to characterize nanomaterials by SRM and the different properties that can be studied (structure, dynamics, composition, ligand number, and distribution, etc.) clearly point toward the standardization of SRM for the characterization of nanomaterials.

SRM also has the ability to give information on nanomaterials biological performance, the nano-bio interactions that will determine its function and thus, its level of efficiency. The life of a non-targeted or targeted nanomedicine, from blood to reaching the target will be discussed, with examples illustrating the use of SRM in each of these steps.

DOI: 10.1201/9781003220688-11

11.1 INTRODUCTION TO NANOMEDICINE

Nanomedicine is the broad field in which nanotechnology is applied to improve the needs of medicine. Still a relatively young field, it was born in early 2000s, and it comprises the fields of regenerative medicine, disease prevention, drug delivery, and diagnostics [1, 2].

By using materials at the nanoscale, different physicochemical properties arise when compared with bulk materials, which allow for applications in sensing, imaging, drug delivery, and therapy. Also, nano is the scale in which many biological processes take place, allowing potential crossing of natural barriers and access to new sites, in which they can interact with biomacromolecules about the same size. Indeed, it is crucial to study the nano-bio interaction in blood and within organs, tissues, and cells to achieve the desired application.

Another aspect that stands out in nanomaterials when compared with traditional, low molecular weight drugs is their superior pharmacokinetic profile: they show a longer blood circulation time, which allows for a better accumulation at the target site. This passive targeting to the disease site can be further enhanced if the nanomaterial is decorated with ligands (antibodies, peptides, aptamers, small molecules) that target the diseased cells, a strategy called active targeting.

Also, nanocarriers stand out for their ability to circumvent physiological barriers like the immune system and the enzymatic and mechanical degradation, and for avoiding renal clearance. Drugs incorporated in nanocarriers remain protected and thus can be active for an extended period of time.

Today, hundreds of nanomedicines are under clinical trials for a wide range of diseases from inflammatory to cardiovascular and neurodegenerative diseases [3–5]. Around 40 products to treat different diseases are currently on the market, mostly for cancer, but also for fungal infections, ocular and blood disorders, among others [6]. If at the beginning mostly liposomal formulations were approved, now a wide variety of types of nanocarriers are on the market: polymeric, nanocrystals, and inorganic nanoparticles. The example of nanoparticle-based Covid-19 vaccines' development has been notorious [7, 8].

11.2 CHALLENGES IN NANOMEDICINE

As nanomedicine is still a relatively new and young field, it is still facing many challenges to accelerate clinical translation. Challenges may change from formulation to formulation, depending on the disease and on the final target site that needs to be reached, but common problems for nano-medicine development can be clearly found. They can be divided into bio-logical and technological challenges, the first one comprising difficulty in biodistribution modulation, crossing biological barriers, heterogeneity in human disease, lack of relevant animal models, and safety and toxicity issues [9, 10]. As for the technological challenges, scaling up nanomedi-cine formulations, lack of specific regulatory guidelines, or cost-benefit considerations are the main issues [9, 10].

Another problem is the unexpected lack of selectivity, as only 0.7% (median) of the administered nanomedicines seem to reach the target site [11].

Finally, the clear lack of standard protocols to characterize nanomateri-als at the physicochemical and physiological/biological level and also for toxicity testing has hampered the development of nanomedicines, con-tributing to the failure in late-stage clinical trials.

Also, the wide variety of materials used caused difficulties in the stan-dardization of the protocols and techniques used for their characteriza-tion. All these problems have led to a poor translation from the bench to the clinic.

After the initial years of nanomedicine, when the expectations were high and probably unrealistic, the initial hype has led to a more prag-matic, mature phase in which collective efforts have to be made in order to address the challenges described above as a research community [12–15]. One of these efforts has been to standardize the characterization and bio-nano interaction of nanomaterials. This standardization has been called MIRIBEL (Minimum Information Reporting in Bio–Nano Experimental Literature) and aims to provide a guide for reproducibility and reliability for the scientific community [16].

11.3 HOW SUPER RESOLUTION CAN HELP NANOMEDICINE

In view of the challenges that nanomedicine is facing, and the clear need for the standardization of nanomaterials characterization and proto-cols for establishing nano-bio interactions, super resolution microscopy (SRM) could be a technique fulfilling some of the gaps and needs of nano-medicine. Super resolution microscopy comprises three main different families: structured-illumination microscopy (SIM), stimulated emission depletion microscopy (STED), and single molecule localization micros-copy (SMLM). How the different techniques work has been described thoroughly in Chapters 2–5 as well as in the recent literature [17–19]. Depending on the scientific question and the type of sample, a certain microscopy method will have to be chosen.

Thanks to reaching the nanoscale, super resolution microscopy is ideal to study nanomaterials, from characterizing them *in vitro* to learning details about their performance in the biological context (nano-bio interaction). Many examples can be found in the literature that will be summarized in the next sections. Recently it has been proposed that SRM could become a standard technique to study nanomaterials [20]. Compared with other nanoscopies, like electron microscopy or atomic force microscopy, SRM allows for an easier, less invasive sample preparation (no drying or freezing is needed), allowing for an observation closer to the native conditions, and not limited to the sample's surface. Besides, it allows for multicolour imaging, has 3D capability, and allows for quantitative molecular analysis and study of real-time dynamics.

11.3.1 Nanomaterials In Vitro Characterization by SRM

Once the material is formulated, certain properties must be robustly described. Characterization is needed for reproducibility issues and also to learn about how those properties link with the biological performance. Mainly, the properties that need to be characterized are as follows: size and shape, size dispersity and aggregation, zeta potential, composition, density, drug loading and release, targeting and labeling [16]. SRM is ideally positioned to elucidate many of those properties, with the possibility of giving such information at the single particle and single molecule level.

As the resolution can go down to a few nanometers, SRM has established the **size and structure** of many different types of nanomaterials, from liposomes to polymeric nanoparticles or peptidic based materials [19]. Examples include peptide assemblies, like diphenylalanine, [21] hydrogels, [22] liposomes, [23] nanostructured lipid carriers (NLC), [24] supramolecular polymers, [25, 26] nanofibers, [27, 28] polymer-based nanomaterials, [29–32] or DNA origami (see Figure 11.1a.) [33, 34]. This information can even be obtained in organic solvents [35]. Also, as information at the single particle level is obtained, size dispersity histograms can be provided to learn about the nanomaterial heterogeneity.

As for the **composition**, valuable information about multicomponent systems can be obtained thanks to the multicolour capability of SRM. Each of the components can be labeled with a spectrally different fluorophore,

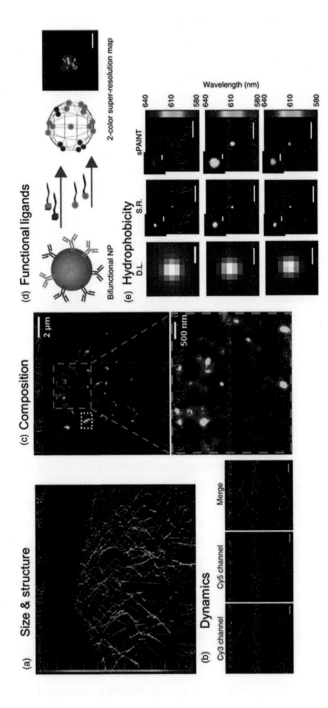

FIGURE 11.1 Nanomaterials properties *in vitro*. (a) Structure of Fmoc-FF hydrogels visualized by PAINT microscopy. (b) Dynamics: STORM imaging of Cy5- and Cy3-labeled BTA polymers (5% labeling) at 1 hour mixing time point. Scale bars, 1 mm. (c) Distribution of dopamine inside DACaPMFs. (d) Functional ligands: multiplexed DNA-PAINT images of particles coated with anti-digoxigenin and anti-biotin antibodies (scale bar 300 nm). (e) Nile Red sPAINT on synthetic 100 nm unilamellar vesicles (LUVs). From top-to-bottom, LUVs composed of DOPC lipid, SM lipid, or SM/cholesterol lipid. Columns from left-to-right; diffraction-limited image (D.L.), super-resolution image (S.R.), sPAINT hydrophobicity map. Scales bars are 500 nm and 20 nm in zoom. ([a] Adapted with permission from [79]; [d] Adapted with permission from [37]; [c] Adapted with permission from [22]; [d] Adapted with permission from [80].)

and their self-sorting in the nanomaterial can be followed in time. Hamachi et al. used STED microscopy to follow the self-sorting of two components of nanofiber hydrogels, a peptide gelator and an amphiphilic phosphate [27]. With a resolution of 80 nm they could follow *in situ*, by time-lapse imaging, the formation of two types of fibers, which showed two different formation rates.

SRM was also used to study the molecular composition of polyplexes. Polyplexes are promising materials for gene therapy, made from the non-covalent complexation between a cationic polymer and a negatively charged oligonucleotide. 2-color quantitative STORM was used to study the effect of a peptide:oligonucleotide ratio on polyplex structure and composition [36].

Also, SRM has been used to characterize the composition of complex organic-inorganic hybrid materials, like dopamine/calcium phosphate organic–inorganic hybrids (DACaPMFs). STORM was used to demonstrate the distribution of the organic component, dopamine, in these hybrids after a one-pot biomineralization process (see Figure 11.1c.) [37].

Moreover, as single particles are visualized, SRM has the potential to also enlighten about the **aggregation state** of the materials, an important parameter for nanomaterial characterization [16].

Remarkably, SRM has the potential to go beyond structural features to obtain data about more complex properties. By following different components, labeled with different fluorophores at different times, one can learn about **dynamic processes**, like the assembly process in supramolecular polymers. Meijer et al. used STORM to study the exchange pathway in BTA-based supramolecular fibers, learning that the interchange is performed along the fiber backbone (see Figure 11.1b) [25]. Peptide amphiphile (PA) nanofibers were also studied using of FRET spectroscopy and STORM [38]. Again, researchers found that the exchange was taking place by monomers or small clusters inserting randomly into fibers, with some areas being fully dynamic, while others were kinetically inactive. Another example is protein nanofibrils, examined combining AFM and STORM, gaining insights about their growth dynamics, exchange kinetics and polymerization mechanism [28].

SRM also has the potential to **count functional molecules** on the surface of nanoparticles. For instance, it is crucial to understand how targeting ligands are distributed and oriented in nanoparticles for an efficient active targeting strategy. So far, only ensemble techniques were used to determine such information, but with DNA-PAINT microscopy it can be obtained at the single particle level with nanometric resolution (see Figure 11.1d) [39, 40]. Moreover, by using DNA-PAINT, the 3D distribution of the functional ligands can be obtained with multiple colors, e.g., one for each type of ligand. Similar information was obtained on non-synthetic systems, virus-like particles (VLPs), by mapping Influenza recombinant proteins expression [41]. The amount of three different Influenza proteins, hemagglutinin (HA), neuraminidase (NA) and ion channel matrix protein 2 (M2), was quantified per single VLP using DNA-PAINT microscopy.

In another study, Archontakis et al. used spectrally resolved STORM (SR-dSTORM) to learn about the antibody orientation on silica nanoparticles [42]. By using two different dyes, one for labeling the antibodies and another for labeling the receptor to which the Cetuximab antibody binds (EGFR), researchers were able to discern between the total amount of antibody and functional antibody, or, what is the same, between antibodies in all kinds of orientations and antibodies that are well-oriented. Information about the number, distribution, and orientation of antibodies on nanoparticles is crucial for the rational design of nanomaterials.

The counting potential of SMLM could also be applied to gain insight on the cargo content per nanoparticle, although examples cannot be found yet, most probably because of the difficulty in accessing the NPs core.

Super resolution microscopy is not limited to just positioning molecules at high accuracy, when adding multi-dimensionality to the parameters read at the microscope, other properties can be obtained. For instance, when using solvatochromic dyes, both the emission spectrum and the position of each emitter can be obtained, generating a map of not only the location, but also the **hydrophobicity** of biological or synthetic structures [43]. By using Nile Red as a probe and spectrally-resolved PAINT (sPAINT), hydrophobicity maps of liposomes were generated, allowing visible changes in hydrophobicity depending on the lipidic composition (see Figure 11.1e). Other properties could be evaluated similarly in the future, like viscosity or surface charge.

Moreover, correlative approaches, in which two different microscopies are used to observe the same sample, offer the possibility to connect two different properties at the single particle level. Recently, this multiparametric analysis has been achieved by correlating DNA-PAINT with TEM, thus gaining dual information on the size and functional number of ligands per NP [44].

The amount of work done to characterize nanomaterials by SRM and the different properties that can be studied (structure, dynamics, composition, ligand number, and distribution, etc.) clearly points toward the standardization of SRM for the characterization of nanomaterials.

11.3.2 SRM to Study Nano-Bio Interactions

As explained, SRM has the ability not only to characterize the nanomaterials *in vitro*, but also to give information on its biological performance, the nano-bio interactions that will determine its function and thus, its level of efficiency [45].

The life of a non-targeted or targeted nanomedicine, from blood to reaching the target will be discussed in this section, with examples illustrating the use of SRM in each of these steps.

One of the first challenges that nanomaterials face when injected in the blood stream is the **interaction with serum proteins**, that might destabilize them or form what is named "protein corona", a layer of adsorbed proteins on the surface of the nanoparticle. Feiner et al. have studied the destabilization of polyplexes in blood serum, by tracking their composition (peptide and mRNA content) and size at different times upon the addition of serum [36]. Although mostly proteomic techniques are used to follow the formation of protein corona composition, SRM could become a complementary tool by giving this information at the single particle level and without much processing of the sample.

Feiner et al. followed the formation, composition, and temporal evolution of protein corona on silica nanoparticles by STORM (see Figure 11.2a) [46]. Interestingly, they found high heterogeneity in protein adsorption and changes, depending on the surface chemistry of the nanoparticle formulation. Similarly, Battaglini et al. studied, by means of a combination of STORM and proteomics, the formation and evolution of protein corona on nanostructured lipid carriers loaded with superparamagnetic iron oxide Nps [47]. Another study, also by STORM, focused on the spatial arrangement of the corona on porous nanoparticles, learning about the penetration depth, depending on the pore size [48]. Wang et al. also studied the protein internalization on porous particles by STED, discerning between core or surface binding [49]. The multicolor ability of STED was exploited to study three different proteins simultaneously, learning that the composition of the corona was influenced by the particle morphology (shape and surface roughness) rather than the size.

After blood circulation, most of the nanocarriers will have to **cross barriers** (endothelial, dermal, blood-brain barrier, intestinal, etc.) to reach their target. Although studies are still scarce on the topic, SRM can be

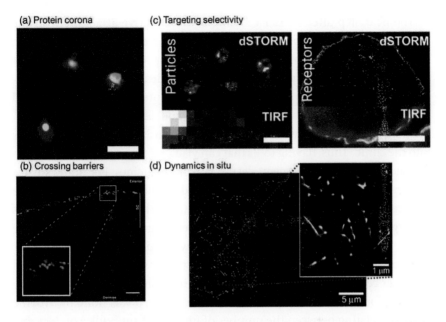

FIGURE 11.2 Nano-bio characterization, part 1. (a) Protein corona: STORM imaging of protein corona formation in mesoporous silica nanoparticles. Nanoparticles are shown in green and the labeled protein in red. Scale bar: 400 nm. (b) Crossing barriers: Cryo-embedded and -sectioned skin exposed to POPC LUVs on the skin surface. LUVs are rarely found inside the SC. Scale bars are 2 µm. (c) Targeting selectivity: dSTORM was used to image and quantify functional Cetuximab on NPs and EGFR on cells at a single-molecule level. A conventional TIRF image is shown for comparison. Scale bar particles 400 nm and receptors 10 µm. (d) Dynamics in situ: dSTORM image of in situ generated nanofibers in a fixed HeLa cell. ([a] Adapted with permission from [81]; [b] Adapted with permission from [50]; [c] Adapted with permission from [82]; [d] Adapted with permission from [83].)

used to study how nanocarriers are able to cross biological barriers to access the disease site. An interesting example is the exploration of transdermal delivery of liposomes on human skin by STED [50]. The authors monitored by STED the delivery of two different liposomes on cryosections of skin, finding that more than crossing the skin, liposomes seem to burst, fuse, and deliver their cargo at the external layers of skin, the one being more flexible, showing a greater delivery (see Figure 11.2b). In view of the growing studies visualizing the endothelial barrier or the BBB by SRM, [51–53] it is just a matter of time until SRM is used to evaluate nanocarriers crossing through this barrier000.

The next step after barrier crossing is the **selective attachment to the target cells**. If nanomedicines may reach around the disease area by passive diffusion, active targeting makes Np interact specifically with the target cells. This is achieved by adding ligands (peptides, aptamers, small molecules, antibodies) on the surface of Nps that will interact with overexpressed receptors on target cells. This step is crucial and, as explained in the introduction, one of the drawbacks for the poor translation of nanomedicines into the clinic. Thus, having tools that can visualize this interaction will help in a more rational and efficient design of nanomedicines. For doing so, it is as important to learn about the number and distribution of ligands on the surface of Nps (see section above) as to quantify the level of cell-receptors on target cells. Few examples on counting receptors on cell surfaces can be found in the literature: from live-PAINT with aptamers to image tumor cell-surface markers (Epidermal growth factor receptor, EGFR) on a panel of cancer cells, [54] PALM imaging of integrin clusters [55] or glyco-PAINT imaging for cell-surface glycans [56]. A further step was done by Woythe et al., correlating between the number of functional antibodies on Nps and the density of receptors on target cells by STORM (see Figure 11.2c.) [57]. Quantifying the number and distribution of Cetuximab antibodies on NPs and EGFR on different cell lines was used to study target-ligand interactions, understanding that, more than the number, the key factor is the ligand distribution for having an optimal targeting activity.

SRM can also be used to visualize the **dynamics of nanomaterials *in situ***. Yao et al. were able to detect the dynamic formation of supramolecular assemblies *in situ*, inside cancer cells, using STORM with a resolution below 50 nm, in an assembly instructed by phosphatase (see Figure 11.2d) [58].

SRM has also been used to study **nanoparticle-cell membrane interactions**, [45] learning about NP internalization in cells (cytoplasmatic barrier crossing) and the uptake mechanisms used to reach intracellular targets.

SRM techniques are able to follow the **trafficking of Nps inside the cell** with high spatio-temporal resolution. STED was used to calculate the total number of internalized NPs by one single cell, applying image segmentation of 3D stacks of whole A459 cells [59]. Van der Zwaag et al. used multicolor STORM to visualize the internalization of polystyrene NPs [60]. Combining STORM imaging with single molecule analysis methods, they could learn about the size, number, and positioning of Nps inside cells. Similarly, STORM was used to image engineered hydrogel nanoparticles inside dendritic cells, following from the engulfment of Nps by the cytoplasmic membrane to lipid raft-mediated trafficking and storage at endo/lysosomal compartments (see Figure 11.3a.) [61]. Sun et al. also used STORM to follow the trafficking of silica Nps, comparing the uptake and distribution among three different cell lines [62].

Metallic nanoparticles give rise to significant contrast when imaged by the light they are able to reflect, as shown by Guggenheim et al. by reflectance-SIM (R-SIM), gaining information about the Np uptake route and uptake form (single or clustered NPs) [63].

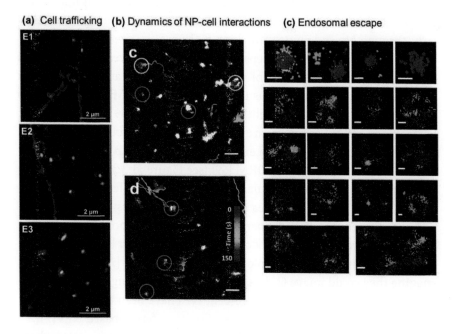

FIGURE 11.3   Nano-bio characterization, part 2. (a) Cell trafficking: In vitro uptake of nanoparticles by dendritic cells. STORM images of PEGPMA nanoparticles internalized by DCs, different stages during the internalization process of the nanoparticles: E1) Engulfment of nanoparticles by the plasma membrane. E2) Intracellular routing and E3) storage in endo/lysosomal vesicles. (b) Dynamics of NP-cell interactions: Comparison of PS40 NPs interacting with CCPs on COS-7 cells: overlay of PS40-PEG(10K)-Tf NP trajectories and CCPs (upper image) and overlay of PS20 NP trajectories and CCPs (bottom image). Scale bars, 1 μm. (c) Endosomal escape: Gallery of STORM images showing ruptured early endosomes and glycoplexes protruding throughout the lipid membrane. For comparison purposes, images representing colocalized and non-colocalized endosomes are shown (first row). The green signal indicates dual-labeled BG-EDA, whereas the red signal denotes early endosomes. Scale bar: 100 nm. ([a] Adapted with permission from [61]; [b] Adapted with permission from [64]; [c] Adapted with permission from [84].)

Dynamics of NP-cell interaction at high spatiotemporal resolution was achieved by combining SRM (PALM to visualize clathrin-coated pits) with single particle tracking (SPT, to trace the NP) (see Figure 11.3b) [64].

Learning about the location of NPs inside the cell gives clues about the **uptake mechanism** and the final fate of NPs. For instance, the multicolour capability of SIM was used to localize metal-organic frameworks (MOFs) in different intracellular vesicles (early endosomes, lysosomes) [65].

Also, to gain insights into endosome maturation, Lee et al. have developed a super-resolution enhanced nanosensor (an ultrasmall fluorescent core-shell aluminosilicate nanoparticle) able to sense the pH intracellularly, through Np endocytic trafficking [66].

One of the more critical steps in intracellular delivery is **endosomal escape**, especially in the case of nucleic acid delivery, as it should reach intact the cytosol or the nucleus to exert its action. SRM has proven very recently its potential to bring light to this difficult to prove phenomena [67]. For instance, mRNA escape from endosomal recycling tubules was observed by SMLM [68] or the rupture of endosome membranes to release complexated-siRNA was imaged by STORM [69] (see Figure 11.3c) or SIM [70].

To learn about the intracellular fate of NPs, SRM offers the possibility of labeling of different organelles in live cells with photoswitchable probes, [71] also in neurons [72]. Lemenager et al. have proposed carbon-dots (C-dots) as intracellular labels of cell organelles due their biocompatibility, small size (5 nm), and multicolor photoluminescence [73]. Using STED imaging they could image C-dots both in fixed and living cells at a resolution of 30 nm. Protein-based fluorescent NPs have been also proposed as probes for SRM, by labeling with Atto647N transferrin-based NPs [74].

It is important information in drug delivery to know when and where the **payload is released** inside the cell. In the case of polyplexes, the decomplexation between the polymer and the nucleic acid and posterior nuclear entry of the payload (a DNA plasmid) was followed by two-color STORM [75, 76]. The intracellular **deformation** of polymeric capsules could also be observed using SIM [77].

Again, as for the characterization step, a correlative approach can be taken to gain multiparametric information about the nano-bio interactions [78].

11.4 FUTURE PERSPECTIVES

The applications of SRM to nanomedicine will grow in the following years, expanding the current applications to others that will further consolidate

SRM as a routine technique for nanomedicine. The assembly state and degradation of nanocarriers, monitoring drug release or nanotoxicology are other aspects that could be studied by SRM. The use of SRM for more complex samples, like tissues, organ-on-a-chip, or even in *in vivo* specimens will further increase the use of SRM.

11.5 CONCLUSIONS

In the field of nanomedicine, the puzzle of how structure-function relationship regarding nanocarriers properties (size, charge, composition, shape, surface functionalization, etc.) still affects their biological performance is to be solved. Super resolution microscopy, with the ability to image nanocarriers both *in vitro* and in the biological context, can greatly help in such a task. Not only the resolution at the nanoscale, but its multicolor and live imaging capability make it an ideal technique to track nanomaterials along their biological journey, linking material properties with its function, and leading to the establishment of structure-activity relationship.

ACKNOWLEDGEMENTS

S.P. acknowledges financial support from the Spanish Ministry of Science and Innovation (PID2019-109450RB-I00/AEI/10.13039/501100011033).

REFERENCES

1. "What is nanomedicine? | ETPN," (2022).
2. C. Domingues, A. Santos, C. Alvarez-Lorenzo, A. Concheiro, I. Jarak, F. Veiga, I. Barbosa, M. Dourado, and A. Figueiras, "Where is nano today and where is it headed? A review of nanomedicine and the dilemma of nanotoxicology," *ACS Nano* **16**, 9994–10041 (2022).
3. C. L. Ventola, "Progress in nanomedicine: approved and investigational nanodrugs," *Pharm. Ther.* **42**, 742–755 (2017).
4. A. C. Anselmo and S. Mitragotri, "Nanoparticles in the clinic," *Bioeng. Transl. Med.* **1**, 10–29 (2016).
5. A. C. Anselmo and S. Mitragotri, "Nanoparticles in the clinic: an update," *Bioeng. Transl. Med.* **4**, e10143 (2019).
6. V. Gadekar, Y. Borade, S. Kannaujia, K. Rajpoot, N. Anup, V. Tambe, K. Kalia, and R. K. Tekade, "Nanomedicines accessible in the market for clinical interventions," *J. Controlled Release* **330**, 372–397 (2021).
7. L. Schoenmaker, D. Witzigmann, J. A. Kulkarni, R. Verbeke, G. Kersten, W. Jiskoot, and D. J. A. Crommelin, "mRNA-lipid nanoparticle COVID-19 vaccines: structure and stability," *Int. J. Pharm.* **601**, 120586 (2021).

8. M. D. Shin, S. Shukla, Y. H. Chung, V. Beiss, S. K. Chan, O. A. Ortega-Rivera, D. M. Wirth, A. Chen, M. Sack, J. K. Pokorski, and N. F. Steinmetz, "COVID-19 vaccine development and a potential nanomaterial path forward," *Nat. Nanotechnol.* **15**, 646–655 (2020).

9. S. Hua, M. B. C. de Matos, J. M. Metselaar, and G. Storm, "Current trends and challenges in the clinical translation of nanoparticulate nanomedicines: pathways for translational development and commercialization," *Front. Pharmacol.* **9**, (2018).

10. H. He, L. Liu, E. E. Morin, M. Liu, and A. Schwendeman, "Survey of clinical translation of cancer nanomedicines—lessons learned from successes and failures," *Acc. Chem. Res.* **52**, 2445–2461 (2019).

11. S. Wilhelm, A. J. Tavares, Q. Dai, S. Ohta, J. Audet, H. F. Dvorak, and W. C. W. Chan, "Analysis of nanoparticle delivery to tumours," *Nat. Rev. Mater.* **1**, 1–12 (2016).

12. K. Park, "The beginning of the end of the nanomedicine hype," *J. Control. Release* **305**, 221–222 (2019).

13. T. J. Webster, "Nanomedicine: real commercial potential or just hype?," *Int. J. Nanomedicine* **1**, 373–374 (2006).

14. J. P. Martins, J. das Neves, M. de la Fuente, C. Celia, H. Florindo, N. Günday-Türeli, A. Popat, J. L. Santos, F. Sousa, R. Schmid, J. Wolfram, B. Sarmento, and H. A. Santos, "The solid progress of nanomedicine," *Drug Deliv. Transl. Res.* **10**, 726–729 (2020).

15. P. Couvreur, "Nanomedicine: from where are we coming and where are we going?," *J. Control. Release* **311–312**, 319–321 (2019).

16. M. Faria, M. Björnmalm, K. J. Thurecht, S. J. Kent, R. G. Parton, M. Kavallaris, A. P. R. Johnston, J. J. Gooding, S. R. Corrie, B. J. Boyd, P. Thordarson, A. K. Whittaker, M. M. Stevens, C. A. Prestidge, C. J. H. Porter, W. J. Parak, T. P. Davis, E. J. Crampin, and F. Caruso, "Minimum information reporting in bio–nano experimental literature," *Nat. Nanotechnol.* **13**, 777–785 (2018).

17. L. Schermelleh, R. Heintzmann, and H. Leonhardt, "A guide to super-resolution fluorescence microscopy," *J. Cell Biol.* **190**, 165–175 (2010).

18. B. Huang, M. Bates, and X. Zhuang, "Super-resolution fluorescence microscopy," *Annu. Rev. Biochem.* **78**, 993–1016 (2009).

19. S. Pujals, N. Feiner-Gracia, P. Delcanale, I. Voets, and L. Albertazzi, "Super-resolution microscopy as a powerful tool to study complex synthetic materials," *Nat. Rev. Chem.* **3**, 68–84 (2019).

20. S. Dhiman, T. Andrian, B. S. Gonzalez, M. M. E. Tholen, Y. Wang, and L. Albertazzi, "Can super-resolution microscopy become a standard characterization technique for materials chemistry?," *Chem. Sci.* **13**, 2152–2166 (2022).

21. S. Pujals, K. Tao, A. Terradellas, E. Gazit, and L. Albertazzi, "Studying structure and dynamics of self-assembled peptide nanostructures using fluorescence and super resolution microscopy," *Chem. Commun.* **53**, 7294–7297 (2017).

22. E. Fuentes, K. Boháčová, A. M. Fuentes-Caparrós, R. Schweins, E. R. Draper, D. J. Adams, S. Pujals, and L. Albertazzi, "PAINT-ing fluorenylmethoxycarbonyl (Fmoc)-diphenylalanine hydrogels," *Chem. – Eur. J.* **26**, 9869–9873 (2020).

23. A. Sharonov and R. M. Hochstrasser, "Wide-field subdiffraction imaging by accumulated binding of diffusing probes," *Proc. Natl. Acad. Sci. U. S. A.* **103**, 18911–18916 (2006).

24. A. Boreham, P. Volz, D. Peters, C. M. Keck, and U. Alexiev, "Determination of nanostructures and drug distribution in lipid nanoparticles by single molecule microscopy," *Eur. J. Pharm. Biopharm. Off. J. Arbeitsgemeinschaft Pharm. Verfahrenstechnik EV* **110**, 31–38 (2017).

25. L. Albertazzi, D. van der Zwaag, C. M. A. Leenders, R. Fitzner, R. W. van der Hofstad, and E. W. Meijer, "Probing exchange pathways in one-dimensional aggregates with super-resolution microscopy," *Science* **344**, 491–495 (2014).

26. S. I. S. Hendrikse, S. P. W. Wijnands, R. P. M. Lafleur, M. J. Pouderoijen, H. M. Janssen, P. Y. W. Dankers, and E. W. Meijer, "Controlling and tuning the dynamic nature of supramolecular polymers in aqueous solutions," *Chem. Commun.* **53**, 2279–2282 (2017).

27. S. Onogi, H. Shigemitsu, T. Yoshii, T. Tanida, M. Ikeda, R. Kubota, and I. Hamachi, "In situ real-time imaging of self-sorted supramolecular nanofibres," *Nat. Chem.* **8**, 743–752 (2016).

28. L. H. Beun, L. Albertazzi, D. van der Zwaag, R. de Vries, and M. A. Cohen Stuart, "Unidirectional living growth of self-assembled protein nanofibrils revealed by super-resolution microscopy," *ACS Nano* **10**, 4973–4980 (2016).

29. C. E. Boott, R. F. Laine, P. Mahou, J. R. Finnegan, E. M. Leitao, S. E. D. Webb, C. F. Kaminski, and I. Manners, "In situ visualization of block copolymer self-assembly in organic media by super-resolution fluorescence microscopy," *Chem. Weinh. Bergstr. Ger.* **21**, 18539–18542 (2015).

30. H. Qiu, Y. Gao, C. E. Boott, O. E. C. Gould, R. L. Harniman, M. J. Miles, S. E. D. Webb, M. A. Winnik, and I. Manners, "Uniform patchy and hollow rectangular platelet micelles from crystallizable polymer blends," *Science* **352**, 697–701 (2016).

31. C. K. Ullal, R. Schmidt, S. W. Hell, and A. Egner, "Block copolymer nanostructures mapped by far-field optics," *Nano Lett.* **9**, 2497–2500 (2009).

32. J. Yan, L.-X. Zhao, C. Li, Z. Hu, G.-F. Zhang, Z.-Q. Chen, T. Chen, Z.-L. Huang, J. Zhu, and M.-Q. Zhu, "Optical nanoimaging for block copolymer self-assembly," *J. Am. Chem. Soc.* **137**, 2436–2439 (2015).

33. C. Steinhauer, R. Jungmann, T. L. Sobey, F. C. Simmel, and P. Tinnefeld, "DNA origami as a nanoscopic ruler for super-resolution microscopy," *Angew. Chem. Int. Ed.* **48**, 8870–8873 (2009).

34. R. Jungmann, M. S. Avendaño, J. B. Woehrstein, M. Dai, W. M. Shih, and P. Yin, "Multiplexed 3D cellular super-resolution imaging with DNA-PAINT and Exchange-PAINT," *Nat. Methods* **11**, 313–318 (2014).

35. A. Aloi, A. Vargas Jentzsch, N. Vilanova, L. Albertazzi, E. W. Meijer, and I. K. Voets, "Imaging nanostructures by single-molecule localization microscopy in organic solvents," *J. Am. Chem. Soc.* **138**, 2953–2956 (2016).

36. N. Feiner-Gracia, R. A. Olea, R. Fitzner, N. El Boujnouni, A. H. van Asbeck, R. Brock, and L. Albertazzi, "Super-resolution imaging of structure, molecular composition, and stability of single oligonucleotide polyplexes," *Nano Lett.* **19**, 2784–2792 (2019).

37. N. Sun, Y. Jia, C. Wang, J. Xia, L. Dai, and J. Li, "Dopamine-mediated biomineralization of calcium phosphate as a strategy to facilely synthesize functionalized hybrids," *J. Phys. Chem. Lett.* **12**, 10235–10241 (2021).

38. R. M. P. da Silva, D. van der Zwaag, L. Albertazzi, S. S. Lee, E. W. Meijer, and S. I. Stupp, "Super-resolution microscopy reveals structural diversity in molecular exchange among peptide amphiphile nanofibres," *Nat. Commun.* **7**, 11561 (2016).

39. P. Delcanale, B. Miret-Ontiveros, M. Arista-Romero, S. Pujals, and L. Albertazzi, "Nanoscale mapping functional sites on nanoparticles by points accumulation for imaging in nanoscale topography (PAINT)," *ACS Nano* **12**, 7629–7637 (2018).

40. T. Andrian, S. Pujals, and L. Albertazzi, "Quantifying the effect of PEG architecture on nanoparticle ligand availability using DNA-PAINT," *Nanoscale Adv.* **3**, 6876–6881 (2021).

41. M. Arista-Romero, P. Delcanale, S. Pujals, and L. Albertazzi, "Nanoscale mapping of recombinant viral proteins: from cells to virus-like particles," *ACS Photonics* **9**, 101–109 (2022).

42. E. Archontakis, L. Woythe, B. van Hoof, and L. Albertazzi, "Mapping the relationship between total and functional antibodies conjugated to nanoparticles with spectrally-resolved direct stochastic optical reconstruction microscopy (SR-dSTORM)," *Nanoscale Adv.* **4**, 4402–4409 (2022).

43. M. N. Bongiovanni, J. Godet, M. H. Horrocks, L. Tosatto, A. R. Carr, D. C. Wirthensohn, R. T. Ranasinghe, J.-E. Lee, A. Ponjavic, J. V. Fritz, C. M. Dobson, D. Klenerman, and S. F. Lee, "Multi-dimensional super-resolution imaging enables surface hydrophobicity mapping," *Nat. Commun.* **7**, 13544 (2016).

44. T. Andrian, P. Delcanale, S. Pujals, and L. Albertazzi, "Correlating super-resolution microscopy and transmission electron microscopy reveals multi-parametric heterogeneity in nanoparticles," *Nano Lett.* **21**, 5360–5368 (2021).

45. X. Chen, Y. Wang, X. Zhang, and C. Liu, "Advances in super-resolution fluorescence microscopy for the study of nano–cell interactions," *Biomater. Sci.* **9**, 5484–5496 (2021).

46. N. Feiner-Gracia, M. Beck, S. Pujals, S. Tosi, T. Mandal, C. Buske, M. Linden, and L. Albertazzi, "Super-resolution microscopy unveils dynamic heterogeneities in nanoparticle protein corona," *Small* **13**, 1701631 (2017).

47. M. Battaglini, N. Feiner, C. Tapeinos, D. D. Pasquale, C. Pucci, A. Marino, M. Bartolucci, A. Petretto, L. Albertazzi, and G. Ciofani, "Combining confocal microscopy, dSTORM, and mass spectroscopy to unveil the evolution of the protein corona associated with nanostructured lipid carriers during blood–brain barrier crossing," *Nanoscale* **14**, 13292–13307 (2022).

48. A. M. Clemments, P. Botella, and C. C. Landry, "Spatial mapping of protein adsorption on mesoporous silica nanoparticles by stochastic optical reconstruction microscopy," *J. Am. Chem. Soc.* **139**, 3978–3981 (2017).

49. Y. Wang, P. E. D. Soto Rodriguez, L. Woythe, S. Sánchez, J. Samitier, P. Zijlstra, and L. Albertazzi, "Multicolor super-resolution microscopy of protein corona on single nanoparticles," *ACS Appl. Mater. Interfaces* **14**, 37345–37355 (2022).

50. J. Dreier, J. A. Sørensen, and J. R. Brewer, "Superresolution and fluorescence dynamics evidence reveal that intact liposomes do not cross the human skin barrier," *PloS One* **11**, e0146514 (2016).
51. E. Sasson, S. Anzi, B. Bell, O. Yakovian, M. Zorsky, U. Deutsch, B. Engelhardt, E. Sherman, G. Vatine, R. Dzikowski, and A. Ben-Zvi, "Nano-scale architecture of blood-brain barrier tight-junctions," *eLife* **10**, e63253 (2021).
52. J. Schlegel, S. Peters, S. Doose, A. Schubert-Unkmeir, and M. Sauer, "Super-resolution microscopy reveals local accumulation of plasma membrane gangliosides at *Neisseria meningitidis* invasion sites," *Front. Cell Dev. Biol.* **7**, 194 (2019).
53. H. Gonschior, V. Haucke, and M. Lehmann, "Super-resolution imaging of tight and adherens junctions: challenges and open questions," *Int. J. Mol. Sci.* **21**, E744 (2020).
54. P. Delcanale, D. Porciani, S. Pujals, A. Jurkevich, A. Chetrusca, K. D. Tawiah, D. H. Burke, and L. Albertazzi, "Aptamers with tunable affinity enable single-molecule tracking and localization of membrane receptors on living cancer cells," *Angew. Chem. Int. Ed.* **59**, 18546–18555 (2020).
55. R. Changede, X. Xu, F. Margadant, and M. P. Sheetz, "Nascent integrin adhesions form on all matrix rigidities after integrin activation," *Dev. Cell* **35**, 614–621 (2015).
56. R. Riera, T. P. Hogervorst, W. Doelman, Y. Ni, S. Pujals, E. Bolli, J. D. C. Codée, S. I. van Kasteren, and L. Albertazzi, "Single-molecule imaging of glycan-lectin interactions on cells with Glyco-PAINT," *Nat. Chem. Biol.* **17**, 1281–1288 (2021).
57. L. Woythe, P. Madhikar, N. Feiner-Gracia, C. Storm, and L. Albertazzi, "A single-molecule view at nanoparticle targeting selectivity: correlating ligand functionality and cell receptor density," *ACS Nano* **16**, 3785–3796 (2022).
58. Q. Yao, C. Wang, M. Fu, L. Dai, J. Li, and Y. Gao, "Dynamic detection of active enzyme instructed supramolecular assemblies *in situ* via super-resolution microscopy," *ACS Nano* **14**, 4882–4889 (2020).
59. H. Peuschel, T. Ruckelshausen, C. Cavelius, and A. Kraegeloh, "Quantification of internalized silica nanoparticles via STED microscopy," *BioMed Res. Int.* 2015, e961208 (2015).
60. D. van der Zwaag, N. Vanparijs, S. Wijnands, R. De Rycke, B. G. De Geest, and L. Albertazzi, "Super resolution imaging of nanoparticles cellular uptake and trafficking," *ACS Appl. Mater. Interfaces* **8**, 6391–6399 (2016).
61. S. De Koker, J. Cui, N. Vanparijs, L. Albertazzi, J. Grooten, F. Caruso, and B. G. De Geest, "Engineering polymer hydrogel nanoparticles for lymph node-targeted delivery," *Angew. Chem. Int. Ed.* **55**, 1334–1339 (2016).
62. N. Sun, Y. Jia, C. Wang, J. Xia, H. Cao, L. Dai, C. Li, X. Zhang, and J. Li, "Monitoring the distribution of internalized silica nanoparticles inside cells via direct stochastic optical reconstruction microscopy," *J. Colloid Interface Sci.* **615**, 248–255 (2022).
63. E. J. Guggenheim, A. Khan, J. Pike, L. Chang, I. Lynch, and J. Z. Rappoport, "Comparison of confocal and super-resolution reflectance imaging of metal oxide nanoparticles," *PloS One* **11**, e0159980 (2016).

64. Y. Li, L. Shang, and G. U. Nienhaus, "Super-resolution imaging-based single particle tracking reveals dynamics of nanoparticle internalization by live cells," *Nanoscale* **8**, 7423–7429 (2016).

65. M. H. Teplensky, M. Fantham, P. Li, T. C. Wang, J. P. Mehta, L. J. Young, P. Z. Moghadam, J. T. Hupp, O. K. Farha, C. F. Kaminski, and D. Fairen-Jimenez, "Temperature treatment of highly porous zirconium-containing metal–organic frameworks extends drug delivery release," *J. Am. Chem. Soc.* **139**, 7522–7532 (2017).

66. R. Lee, J. A. Erstling, J. A. Hinckley, D. V. Chapman, and U. B. Wiesner, "Addressing particle compositional heterogeneities in super-resolution-enhanced live-cell ratiometric pH sensing with ultrasmall fluorescent core–shell aluminosilicate nanoparticles," *Adv. Funct. Mater.* **31**, 2106144 (2021).

67. T. Andrian, R. Riera, S. Pujals, and L. Albertazzi, "Nanoscopy for endosomal escape quantification," *Nanoscale Adv.* **3**, 10–23 (2021).

68. P. Paramasivam, C. Franke, M. Stöter, A. Höijer, S. Bartesaghi, A. Sabirsh, L. Lindfors, M. Y. Arteta, A. Dahlén, A. Bak, S. Andersson, Y. Kalaidzidis, M. Bickle, and M. Zerial, "Endosomal escape of delivered mRNA from endosomal recycling tubules visualized at the nanoscale," *J. Cell Biol.* **221**, e202110137 (2021).

69. M. Wojnilowicz, A. Glab, A. Bertucci, F. Caruso, and F. Cavalieri, "Super-resolution imaging of proton sponge-triggered rupture of endosomes and cytosolic release of small interfering RNA," *ACS Nano* **13**, 187–202 (2019).

70. S. Ben Djemaa, K. Hervé-Aubert, L. Lajoie, A. Falanga, S. Galdiero, S. Nedellec, M. Soucé, E. Munnier, I. Chourpa, S. David, and E. Allard-Vannier, "gH625 cell-penetrating peptide promotes the endosomal escape of nanovectorized siRNA in a triple-negative breast cancer cell line," *Biomacromolecules* **20**, 3076–3086 (2019).

71. S.-H. Shim, C. Xia, G. Zhong, H. P. Babcock, J. C. Vaughan, B. Huang, X. Wang, C. Xu, G.-Q. Bi, and X. Zhuang, "Super-resolution fluorescence imaging of organelles in live cells with photoswitchable membrane probes," *Proc. Natl. Acad. Sci. U. S. A.* **109**, 13978–13983 (2012).

72. C. Werner, M. Sauer, and C. Geis, "Super-resolving microscopy in neuroscience," *Chem. Rev.* **121**, 11971–12015 (2021).

73. G. Leménager, E. D. Luca, Y.-P. Sun, and P. P. Pompa, "Super-resolution fluorescence imaging of biocompatible carbon dots," *Nanoscale* **6**, 8617–8623 (2014).

74. L. Shang, P. Gao, H. Wang, R. Popescu, D. Gerthsen, and G. U. Nienhaus, "Protein-based fluorescent nanoparticles for super-resolution STED imaging of live cells," *Chem. Sci.* **8**, 2396–2400 (2017).

75. R. Riera, N. Feiner-Gracia, C. Fornaguera, A. Cascante, S. Borrós, and L. Albertazzi, "Tracking the DNA complexation state of pBAE polyplexes in cells with super resolution microscopy," *Nanoscale* **11**, 17869–17877 (2019).

76. R. Riera, J. Tauler, N. Feiner-Gracia, S. Borrós, C. Fornaguera, and L. Albertazzi, "Complex pBAE nanoparticle cell trafficking: tracking both position and composition using super resolution microscopy," *ChemMedChem* **17**, e202100633 (2022).

77. X. Chen, J Cui, H. Sun, M. Müllner, Y. Yan, K. F. Noi, Y. Ping, and F. Caruso, "Analysing intracellular deformation of polymer capsules using structured illumination microscopy," *Nanoscale* **8**, 11924–11931 (2016).

78. J. Groen, A. Palanca, A. Aires, J. J. Conesa, D. Maestro, S. Rehbein, M. Harkiolaki, A. V. Villar, A. L. Cortajarena, and E. Pereiro, "Correlative 3D cryo X-ray imaging reveals intracellular location and effect of designed anti-fibrotic protein–nanomaterial hybrids," *Chem. Sci.* **12**, 15090–15103 (2021).

79. P. Delcanale, B. Miret-Ontiveros, M. Arista-Romero, S. Pujals, and L. Albertazzi, "Nanoscale mapping functional sites on nanoparticles by points accumulation for imaging in nanoscale topography (PAINT)," *ACS Nano* **12**, 7629–7637 (2018).

80. M. N. Bongiovanni, J. Godet, M. H. Horrocks, L. Tosatto, A. R. Carr, D. C. Wirthensohn, R. T. Ranasinghe, J.-E. Lee, A. Ponjavic, J. V. Fritz, C. M. Dobson, D. Klenerman, and S. F. Lee, "Multi-dimensional super-resolution imaging enables surface hydrophobicity mapping," *Nat. Commun.* **7**, 13544 (2016).

81. N. Feiner-Gracia, M. Beck, S. Pujals, S. Tosi, T. Mandal, C. Buske, M. Linden, and L. Albertazzi, "Super-resolution microscopy unveils dynamic heterogeneities in nanoparticle protein corona," *Small Weinh. Bergstr. Ger.* **13**, (2017).

82. L. Woythe, P. Madhikar, N. Feiner-Gracia, C. Storm, and L. Albertazzi, "A single-molecule view at nanoparticle targeting selectivity: correlating ligand functionality and cell receptor density," *ACS Nano* **16**, 3785–3796 (2022).

83. Q. Yao, C. Wang, M. Fu, L. Dai, J. Li, and Y. Gao, "Dynamic detection of active enzyme instructed supramolecular assemblies *in situ* via super-resolution microscopy," *ACS Nano* **14**, 4882–4889 (2020).

84. M. Wojnilowicz, A. Glab, A. Bertucci, F. Caruso, and F. Cavalieri, "Super-resolution imaging of proton sponge-triggered rupture of endosomes and cytosolic release of siRNA," *ACS Nano* **13**, 187–202 (2019).

CHAPTER 12

Super-Resolution Microscopy Applications to Catalysis

Ruben F. Hamans and Andrea Baldi

Vrije Universiteit Amsterdam, Amsterdam, The Netherlands

12.1 INTRODUCTION

Catalysts increase the rate of a chemical reaction without being consumed. Their presence in the chemical industry is ubiquitous, with the production of ~90% of all chemicals being assisted by catalyst materials [1]. Ideally, the optimization of these catalytic processes is done using knowledge of the relationship between the performance of the catalyst and its local structure and composition. Traditionally, these relationships are established using characterizations at the ensemble level. Therefore, interpreting data is done under the assumption that all catalyst particles behave similarly, even though structure and composition is known to vary greatly from particle to particle. Clearly then, the characterization of catalyst particles should ideally be performed at the single particle level [2, 3], with a temporal resolution on the order of the turnover rate of the reaction under study and a spatial resolution on the order of the typical length scale of the catalyst (e.g., its size or the size of its pores).

In this regard, single molecule localization microscopy has recently emerged as a powerful tool due to its high temporal (order 10 ms) and spatial resolution (order 10 nm), its non-invasiveness, and its ability to measure in real time and under in-situ conditions [4–9]. This technique has been used to study heterogeneous catalysis in the liquid phase on a variety of catalyst materials, ranging from layered double hydroxides [10],

DOI: 10.1201/9781003220688-12

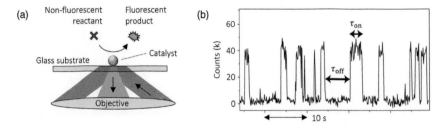

FIGURE 12.1 (a) Illustration of a fluorogenic chemical reaction imaged in a fluorescence microscope. (b) Fluorescence intensity as a function of time of a single ~6 nm Au nanoparticle exposed to 0.05 µM resazurin and 1 mM NH₂OH, which generates the fluorescent product resorufin. ([a] adapted under a Creative Commons CC-BY license from [3]. Copyright 2020 The Authors; [b] Adapted with permission from [11]. Copyright 2008 Springer Nature.)

to metal nanoparticles (Au, Ag, Pd, Pt, Cu) [11–33], zeolites [34–41], semiconductor nanoparticles (TiO$_2$, BiVO$_4$, CdS) [13, 16, 19, 42–52], and carbon nanotubes [53]. In the context of catalysis, single molecule localization microscopy often relies on the use of so-called fluorogenic reactions: chemical reactions in which a non-fluorescent reactant molecule is catalytically converted to a fluorescent product molecule (Figure 12.1a). The single molecule sensitivity can here be achieved by controlling the turnover rate of the reaction, for example by varying the pH or the concentration of the reactant molecule. When the dissociation or photobleaching time of a product molecule (τ_{on}) is smaller than the time in between consecutive reactions (τ_{off}), the typical blinking associated with single molecule localization microscopy is observed (Figure 12.1b).

In this chapter, we will discuss the application of single molecule localization microscopy to heterogeneous catalysis. We have selected several key works that demonstrate the ability of this technique to reveal insights that would otherwise remain hidden in ensemble experiments. The chapter is separated in sections by the catalyst material: layered double hydroxides, metal nanoparticles, zeolites, and semiconductor nanoparticles. Finally, we end this chapter with an outlook on how single molecule localization microscopy can further be used to advance the field of heterogeneous catalysis.

12.2 LAYERED DOUBLE HYDROXIDES

The use of fluorogenic chemical reactions to image catalytic reactions at the single turnover level was first reported in 2006 by Roeffaers et al. on [Li⁺-Al³⁺] layered double hydroxide catalysts [10]. The authors studied the

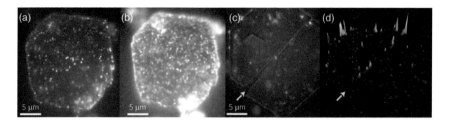

FIGURE 12.2 (a,b) Fluorescence image of the transesterification of C-FDA with 1-butanol at 40 nM (a) and 700 nM (b). (c,d) Fluorescence image (c) and accumulated spot intensity on the same crystal over 256 consecutive images (d) of the hydrolysis of C-FDA at 600 nM. Yellow arrows indicate the same viewing direction on the same crystal. (Adapted with permission from [10]. Copyright 2006 Springer Nature.)

probe molecule 5-carboxyfluorescein diacetate (C-FDA), which becomes emissive upon catalytic transesterification with 1-butanol (Figure 12.2a, b), or upon catalytic hydrolysis in water-containing media (Figure 12.2c, d). They found that transesterification occurs on the entire outer crystal surface (Figure 12.2a, b), while ester hydrolysis proceeds on the lateral $\{10\bar{1}0\}$ crystal faces (Figure 12.2c, d). This information could previously only be derived indirectly, for example by correlating catalytic activity with ex-situ characterizations. Fluorescence microscopy, however, has sufficient resolution to observe the activity of different crystal faces in-situ.

12.3 METAL NANOPARTICLES

In 2008, Xu et al. reported the use of fluorescence microscopy to image the reductive deoxygenation of resazurin (non-fluorescent) to resorufin (highly fluorescent) at the single turnover level on individual ~6 nm Au particles (Figure 12.1b) [11]. Statistical analysis of the product formation and dissociation times (τ_{off} and τ_{on}, Figure 12.1b) revealed a large heterogeneity in catalytic activity, despite the narrow size distribution of the particles (6.0 ± 1.7 nm). Furthermore, by measuring the τ_{on} times as a function of resazurin concentration, the authors found three distinct mechanisms for product dissociation: while most particles preferred dissociation assisted by a reactant binding step, some preferred a direct dissociation of the product, and others showed no preference between these two mechanisms. This diverse behavior demonstrates the heterogeneity in catalytic properties of metal nanoparticles and highlights the importance of single particle approaches to unveil their structure-function relationship.

Statistical analysis of the product formation time τ_{off} has also been used to gain insight in the reaction pathway of amplex red oxidation to resorufin [20]. Reactions occurring within a single step (i.e., without an intermediate) are characterized by an exponential decay in their distribution of τ_{off} times. Conversely, reactions with an intermediate show a peak in their τ_{off} distribution (Figure 12.3a). This distinction can only be made when performing these analyses on individual particles, since the characteristic peak in the τ_{off} distribution is completely washed out upon ensemble averaging, due to the large heterogeneity in catalytic activity (Figure 12.3b).

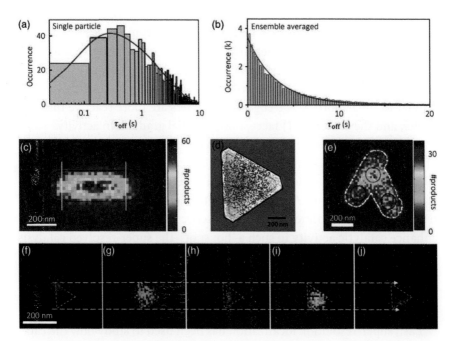

FIGURE 12.3   (a,b) Histogram of the τ_{off} times measured on a single Au nanorod (a) and summed over 90 nanorods (b). (c–e) Fluorescent products detected on an Au nanorod (c), a triangular Au nanoplate (d), a dimer of two Au nanorods with a nanoscale gap (e). The products in the panel (d) are detected on the flat {111} facet (red), the edges (blue), or the corners (green) of the particle. (f–j) Two-dimensional histograms of the products detected on a single triangular Au nanoplate, separated in groups of 1,000 products. ([a,b] Adapted with permission from [20]. Copyright 2014 American Chemical Society. [c] Adapted with permission from [15]. Copyright 2012 Springer Nature. [d] Adapted with permission from [17]. Copyright 2013 American Chemical Society. [e] Adapted with permission from [27]. Copyright 2018 American Chemical Society. [f]–[j] Adapted with permission from [22]. Copyright 2015 National Academy of Sciences.)

Tuning fluorogenic reactions down to single molecule sensitivity not only allows for the analysis of temporal information, but also opens the possibility of performing super-resolution localization, which reveals a wealth of information that would not be accessible when using diffraction-limited techniques. Zhou et al. showed that single Au nanorods are generally more reactive at their tips due to the presence of more corner and edge sites [15]. However, this study again revealed a large heterogeneity, with some particles showing more activity in the center (Figure 12.3c). Furthermore, within the single facet between the tips, the nanorods show a gradient in catalytic activity, from the center to the edges of the particle (see again Figure 12.3c). The authors attributed this observation to an underlying gradient of surface defects resulting from a decay in growth rate during the synthesis of the particles. This behavior also extends to Au nanoplates [17], which can be separated into three regions, as can be seen in Figure 12.3d: flat {111} facet (red), edges (blue), and corners (green). Similar to the nanorods, the edges and corners are more reactive than the center, and the center {111} facet shows a radial gradient in activity.

Super-resolution localization has also been used to study the influence of exciting plasmons in metal nanoparticles on their catalytic activity [19, 27, 31, 32]. Zou et al. studied the reductive deoxygenation of resazurin to resorufin on dimers of Au nanorods with a nanoscale gap [27]. This gap results in strong confinement of the electric field, which can potentially enhance catalytic activity in such a 'hot spot'. The super-resolved catalysis maps show, in fact, that enhanced product formation is observed inside the gap between the nanorods (Figure 12.3e). This enhancement could be attributed to the generation of energetic charge carriers as a result of plasmon decay.

Last, super-resolution localization has been used to study how the catalytic activity of metal nanoparticles vary over time. Zhang et al. showed that the super-resolved map of products detected on a single triangular Au nanoplate over a period of up to ~10 h accurately reproduces the nanoparticle shape [22]. Similar to the studies by Zhou et al. [15] and Andoy et al. [17], the center of the particle shows a linear gradient in catalytic activity due to an underlying gradient of surface defects. Separating the catalytic activity maps into subgroups of 1,000 products, however, also reveals spatiotemporal variations in activity (Figure 12.3f–j). The reactivity of the nanoparticle corners was initially found to be spatially heterogeneous, with one corner being particularly active, but eventually converged to the point where all corners behaved similarly. The authors attributed this dynamic behavior to restructuring of the surface, with the nanoparticle corners transitioning from being sharp to blunt.

The presence of reaction intermediates, the influence of defects, plasmon excitation, and of dynamic restructuring on catalytic activity would have remained hidden in ensemble or diffraction-limited measurements.

12.4 ZEOLITES

Zeolites are porous aluminosilicate minerals used as catalysts for a wide variety of industrial processes, such as fluid catalytic-cracking and the oxidation and epoxidation of hydrocarbons [35, 38]. Despite the macroscopic dimensions of zeolite particles, their catalytic properties are governed by nanoscale compositional and structural features, making single molecule localization microscopy a suitable tool for the in-situ study of these catalytic materials.

Roeffaers et al. used the fluorescent products of furfuryl alcohol oligomerization to study the catalytic activity of ZSM-22 zeolite particles, which are needle-like crystallites with widths below the diffraction limit [34]. Furfuryl alcohol is particularly suitable here, as it is small enough to enter the pores of many heterogeneous catalysts, unlike the large polycyclic molecules that are often used as fluorogenic probes. The catalytic activity maps revealed zones of varying activity within a single rod, which the authors attributed to an aggregation of active sites that occurs during the ZSM-22 growth. The authors also measured on ZSM-5 zeolites, which are micrometer-sized particles that have a complex pore structure. The catalytic-activity maps revealed a very narrow active zone at the boundaries of the crystal, with dimensions well below the diffraction limit (Figure 12.4a), likely due to the influx of reagents from two different crystal faces at these edges. The authors also observed nanometer-sized structural kinks (see again Figure 12.4a), which arise due to structural defects in the crystal.

De Cremer et al. studied the catalytic activity of Ti-MCM-41A particles [35]. Upon epoxidation of the butadienyl bridge of phenylbutadienyl-substituted boron dipyrromethene difluoride (PBD-bodipy), the emission undergoes a strong blue shift of 50–80 nm, allowing for selective spectral filtering of the reaction products. Furthermore, the size of PBD-bodipy is comparable to that of large olefinic substrates such as cholesterol, which are frequently used for epoxidation with Ti-MCM-41. The authors found that catalytic reactions occur predominantly in the outer rim of the particle (Figure 12.4b, c). Complementary XPS measurements show that Ti is

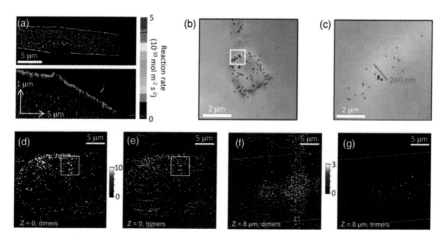

FIGURE 12.4 (a) Map of the reaction rate of furfuryl alcohol oligomerization on a single ZSM-5 zeolite particle. (b,c) Turnovers of the epoxidation of PBD-bodipy on a single Ti-MCM-41 particle. (d–g) Maps of short-lived (d & f, τ_{on} < 0.7 s; dimeric) and long-lived (e & g, τ_{on} > 0.7 s; trimeric) species, measured at the outer surface of a zeolite H-ZSM-5 (d & e) and at z = 8 µm (f & g). The color scale denotes the total number of emitters per 200 × 200 nm² detected over 500 seconds. ([a] Adapted with permission from [34]. Copyright 2009 John Wiley and Sons. [b,c] Adapted with permission from [35]. Copyright 2010 John Wiley and Sons. [d]–[g] Adapted with permission under a Creative Commons CC-BY 4.0 license from [41]. Copyright 2015 American Chemical Society.)

homogeneously distributed across the particle, and sorption experiments show that the pores are fully accessible. Therefore, the rate-limiting process here is intraparticle diffusion (i.e., reactions occur before molecules can reach the inner parts of the particle). These experiments provide mechanistic insight into why small Ti-MCM-41 particles tend to be more reactive, which is an observation that previously was only made phenomenologically from ensemble experiments.

The optimization of the catalytic performance of zeolites often relies on postsynthetic treatments, such as dealumination, which simultaneously reduces the aluminum content and introduces mesoporosity, thereby allowing better diffusion. To avoid a trial-and-error based optimization, single molecule localization microscopy can provide insight at the relevant length scales. Liu et al. used furfuryl alcohol oligomerization to study the catalytic activity of acid mordenites with varying degrees of dealumination [37]. Before dealumination, the particles showed activity mostly at the crystal periphery, due to limited pore accessibility. Conversely, strongly dealuminated particles showed a homogeneous catalytic activity throughout the particle, indicating better pore accessibility. Contrary to the strongly- and non-dealuminated particles, the mildly dealuminated particles showed a pronounced intercrystal heterogeneity and could be divided in three groups: ~33% showed activity only at the outer regions, ~25% showed activity mostly at the core, ~28% showed activity across the whole particle, and the remaining ~14% showed transitional behavior among the aforementioned groups. These remarkable heterogeneities are the result of an interplay between the acid site distribution and molecular accessibility. To image the acid site distribution, the authors measured stimulated Raman scattering of d-acetonitrile, which is small enough to access all acid sites, and of benzonitrile, which can only access larger pores. A homogeneous distribution of acid sites was found for all dealumination levels, with a decreasing density as the dealumination becomes more intense. The variations in catalytic activity are then the result of varying site accessibility for larger molecules, such as furfuryl alcohol. The combination of super-resolution and Raman techniques here give unique insights into the influence of dealumination on pore accessibility, acid site distribution, and, consequently, catalytic activity.

Ristanović et al. used the Brønsted-acid-catalyzed oligomerization of styrene derivatives to study the influence of solvent polarity and defect density on the reactivity of zeolite H-ZSM-5 [41]. When using 4-methoxystyrene as a reactant, a very broad distribution of τ_{on} times was measured, with τ_{on} being particularly high at the crystal surface. The oligomerization reaction can result in the formation of small, linear dimeric carbocations. Further oligomerization results in larger trimeric or cyclic dimeric carbocations, which are more stable and hence have long τ_{on} times, as further oligomerization is unlikely to happen. At the outer surface, where acid sites are more accessible, both short-lived dimers and long-lived trimers are observed, as can be seen in Figure 12.4d,e, where products are separated by their τ_{on} time. In the middle of the zeolite crystal, however, the smaller dimeric species are more dominant due to limited pore accessibility (Figure 12.4f,g). To study the effect of defect density, the authors studied two types of zeolite crystals, the highly microporous parent zeolite H-ZSM-5 and defect-rich, mildly steamed H-ZSM-5 with induced mesoporosity. On the parent crystal, fairly low reactivity was observed due to pore blockage. Conversely, steaming induces defects in the crystal, which was found to greatly enhance its catalytic activity, to the point where the reactant concentration needed to be diluted significantly (50×) to still satisfy the single molecule requirement. Lastly, a significant decrease in activity was observed when the solvent is changed from n-heptane to 1-butanol, as 1-butanol also adsorbs on the acid sites and, therefore, competes with the fluorogenic reaction.

The findings in all the above studies relied on super-resolved catalytic activity maps (Figure 12.4) and would, therefore, not be obvious when using conventional microscopy techniques.

12.5 SEMICONDUCTOR NANOPARTICLES

Semiconductor nanoparticles, particularly TiO_2, have been extensively studied and used as photocatalysts for, for example, water splitting and the degradation of organic pollutants [42, 54]. Photocatalysis on these materials relies on the excitation of charge carriers inside the semiconductor and their subsequent transfer to adsorbate species. Due to the small-length scales involved, such as the electron mean free path, single molecule localization microscopy can here reveal mechanistic insights into the charge transfer processes that would otherwise remain hidden in ensemble experiments [13, 16, 19, 42–52].

Tachikawa et al. have designed a fluorogenic probe that can detect electron transfer from a TiO_2 surface [45, 48]. They have used this probe to characterize the catalytic activity of an anatase TiO_2 crystal, consisting of a {001} center facet and {101} side facets [46]. Upon irradiating the center facet with UV light at 365 nm, which generates charge carriers inside the crystal that can trigger the fluorogenic probe, it was found that a significant number of charges could migrate toward the {101} side facets (Figure 12.5a). When irradiating the side facet, charges stayed accumulated and the reaction proceeds on the same facet (Figure 12.5b). This finding, which would not be evident from bulk measurements, can be attributed to the surface energy levels, with the {101} side facets having a slightly lower conduction band energy.

The same group later found that photocatalysis on TiO_2 can also occur with visible light irradiation at 488 nm when loading the semiconductor with Au nanoparticles [48]. Super-resolution maps showed that the reactive sites were spatially distributed within a distance of a few tens of nanometers from the deposited Au nanoparticles (Figure 12.5c,d). The authors attributed this observation to the plasmon resonance of the Au nanoparticles, which can decay into the electronic defect states of TiO_2, which can subsequently trigger the fluorogenic reaction.

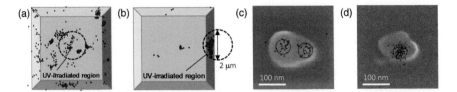

FIGURE 12.5 (a,b) Turnovers of electron transfer from TiO_2 upon UV irradiation of the {001} center facet (a) or the {101} side facet (b). Reactions occurring on the center facet are denoted in blue and those occurring on the side facet in red. (c,d) Turnovers of electron transfer from Au/TiO_2 particles. Dashed circles indicate the reactive sites with Au nanoparticles. ([a,b] Adapted with permission from [46]. Copyright 2011 American Chemical Society. [c,d] adapted with permission from [48]. Copyright 2013 American Chemical Society.)

12.6 CONCLUSION AND OUTLOOK

In this chapter, we have seen how single molecule localization microscopy has been used to characterize catalytic materials in real time and under in-situ conditions. Due to its high temporal and spatial resolution, this technique has revealed mechanistic insights that would remain hidden in ensemble-averaged or diffraction-limited measurements.

An obvious limitation of the studies presented in this chapter is that the catalytic reaction under study needs to be fluorogenic. Most chemical processes, however, do not involve fluorescent species. This limitation has recently been addressed with the development of a technique in which a single nanoparticle catalyzes two reactions simultaneously [51]. The first reaction is the reaction of interest and its reactants and products do not fluoresce. The second auxiliary reaction is fluorogenic and single molecule localization can be applied. If both reactions compete for the same surface sites, the reaction of interest suppresses the rate of the fluorogenic reaction. The extent of suppression can be imaged using super-resolution microscopy, thereby giving spatial information on the reaction of interest [33, 51].

Last, many studies presented in this chapter have highlighted the importance of defects on catalytic activity, with reactions often taking place in regions of high defect density or at corner and edge sites. As these features in the catalyst material ultimately occur at the atomic level, further technological advancements can here reveal more mechanistic insights. Examples include finding super-resolution techniques with increased localization precision, a full implementation of 3D super-resolution microscopy [9, 55], and combining super-resolution microscopy with correlative transmission electron microscopy [56, 57].

REFERENCES

[1] I. Chorkendorff and J. W. Niemantsverdriet, *Concepts of Modern Catalysis and Kinetics* (John Wiley & Sons, 2017).

[2] I. L. Buurmans and B. M. Weckhuysen, "Heterogeneities of individual catalyst particles in space and time as monitored by spectroscopy," *Nat. Chem.* **4**, 873–886 (2012).

[3] R. F. Hamans, R. Kamarudheen, and A. Baldi, "Single particle approaches to plasmon-driven catalysis," *Nanomaterials* **10**, 2377 (2020).

[4] G. De Cremer, B. F. Sels, D. E. De Vos, J. Hofkens, and M. B. Roeffaers, "Fluorescence micro(spectro)scopy as a tool to study catalytic materials in action," *Chem. Soc. Rev.* **39**, 4703–4717 (2010).

[5] K. P. Janssen, G. De Cremer, R. K. Neely, A. V. Kubarev, J. Van Loon, J. A. Martens, D. E. De Vos, M. B. Roeffaers, and J. Hofkens, "Single molecule methods for the study of catalysis: from enzymes to heterogeneous catalysts," *Chem. Soc. Rev.* **43**, 990–1006 (2014).

[6] J. B. Sambur and P. Chen, "Approaches to single-nanoparticle catalysis," *Ann. Rev. Phys. Chem.* **65**, 395–422 (2014).

[7] P. Chen, X. Zhou, N. M. Andoy, K.-S. Han, E. Choudhary, N. Zou, G. Chen, and H. Shen, "Spatiotemporal catalytic dynamics within single nanocatalysts revealed by single-molecule microscopy," *Chem. Soc. Rev.* **43**, 1107–1117 (2014).

[8] T. Chen, B. Dong, K. Chen, F. Zhao, X. Cheng, C. Ma, S. Lee, P. Zhang, S. H. Kang, J. W. Ha et al., "Optical super-resolution imaging of surface reactions," *Chem. Rev.* **117**, 7510–7537 (2017).

[9] W. Y. Teoh, A. Urakawa, Y. H. Ng, and P. Sit, *Heterogeneous Catalysts: Advanced Design, Characteriza- tion, and Applications* (John Wiley & Sons, 2021).

[10] M. B. Roeffaers, B. F. Sels, H. Uji-I, F. C. De Schryver, P. A. Jacobs, D. E. De Vos, and J. Hofkens, "Spatially resolved observation of crystal-face-dependent catalysis by single turnover counting," *Nature* **439**, 572–575 (2006).

[11] W. Xu, J. S. Kong, Y.-T. E. Yeh, and P. Chen, "Single-molecule nanocatalysis reveals heterogeneous reaction pathways and catalytic dynamics," *Nat. Mater.* **7**, 992–996 (2008).

[12] X. Zhou, W. Xu, G. Liu, D. Panda, and P. Chen, "Size-dependent catalytic activity and dynamics of gold nanoparticles at the single-molecule level," *J. Amer. Chem. Soc.* **132**, 138–146 (2010).

[13] N. Wang, T. Tachikawa, and T. Majima, "Single-molecule, single-particle observation of size-dependent photocatalytic activity in au/tio$_2$ nanocomposites," *Chem. Sci.* **2**, 891–900 (2011).

[14] K. S. Han, G. Liu, X. Zhou, R. E. Medina, and P. Chen, "How does a single Pt nanocatalyst behave in two different reactions? A single-molecule study," *Nano Lett.* **12**, 1253–1259 (2012).

[15] X. Zhou, N. M. Andoy, G. Liu, E. Choudhary, K.-S. Han, H. Shen, and P. Chen, "Quantitative super-resolution imaging uncovers reactivity patterns on single nanocatalysts," *Nat. Nanotechn.* **7**, 237– 241 (2012).

[16] Z. Bian, T. Tachikawa, W. Kim, W. Choi, and T. Majima, "Superior electron transport and photocatalytic abilities of metal-nanoparticle-loaded TiO$_2$ superstructures," *J. Phys. Chem. C* **116**, 25444–25453 (2012).

[17] N. M. Andoy, X. Zhou, E. Choudhary, H. Shen, G. Liu, and P. Chen, "Single-molecule catalysis mapping quantifies site-specific activity and uncovers radial activity gradient on single 2D nanocrystals," *J. Amer. Chem. Soc.* **135**, 1845–1852 (2013).

[18] R. Han, J. W. Ha, C. Xiao, Y. Pei, Z. Qi, B. Dong, N. L. Bormann, W. Huang, and N. Fang, "Geometry-assisted three-dimensional superlocalization imaging of single-molecule catalysis on modular multilayer nanocatalysts," *Angew. Chem. Int. Ed.* **53**, 12865–12869 (2014).

[19] J. W. Ha, T. P. A. Ruberu, R. Han, B. Dong, J. Vela, and N. Fang, "Super-resolution mapping of photogenerated electron and hole separation in single metal-semiconductor nanocatalysts," *J. Amer. Chem. Soc.* **136**, 1398–1408 (2014).

[20] H. Shen, X. Zhou, N. Zou, and P. Chen, "Single-molecule kinetics reveals a hidden surface reaction inter- mediate in single-nanoparticle catalysis," *J. Phys. Chem. C* **118**, 26902–26911 (2014).

[21] M. R. Decan and J. C. Scaiano, "Study of single catalytic events at copper-in-charcoal: localization of click activity through subdiffraction observation of single catalytic events," *J. Phys. Chem. Letters* **6**, 4049–4053 (2015).

[22] Y. Zhang, J. M. Lucas, P. Song, B. Beberwyck, Q. Fu, W. Xu, and A. P. Alivisatos, "Superresolution fluorescence mapping of single-nanoparticle catalysts reveals spatiotemporal variations in surface reactivity," *Proc. Natl. Acad. Sci. U.S.A.* **112**, 8959–8964 (2015).

[23] Y. Zhang, T. Chen, S. Alia, B. S. Pivovar, and W. Xu, "Single-molecule nanocatalysis shows in situ deactivation of Pt/C electrocatalysts during the hydrogen-oxidation reaction," *Angew. Chem. Int. Ed.* **55**, 3086–3090 (2016).

[24] T. Chen, Y. Zhang, and W. Xu, "Single-molecule nanocatalysis reveals catalytic activation energy of single nanocatalysts," *J. Amer. Chem. Soc.* **138**, 12414–12421 (2016).

[25] T. Chen, S. Chen, P. Song, Y. Zhang, H. Su, W. Xu, and J. Zeng, "Single-molecule nanocatalysis reveals facet-dependent catalytic kinetics and dynamics of palladium nanoparticles," *ACS Catal.* **7**, 2967–2972 (2017).

[26] G. Chen, N. Zou, B. Chen, J. B. Sambur, E. Choudhary, and P. Chen, "Bimetallic effect of single nanocatalysts visualized by super-resolution catalysis imaging," *ACS Cent. Sci.* **3**, 1189–1197 (2017).

[27] N. Zou, G. Chen, X. Mao, H. Shen, E. Choudhary, X. Zhou, and P. Chen, "Imaging catalytic hotspots on single plasmonic nanostructures via correlated super-resolution and electron microscopy," *ACS Nano* **12**, 5570–5579 (2018).

[28] B. Dong, Y. Pei, F. Zhao, T. W. Goh, Z. Qi, C. Xiao, K. Chen, W. Huang, and N. Fang, "In situ quantitative single-molecule study of dynamic catalytic processes in nanoconfinement," *Nat. Cat.* **1**, 135–140 (2018).

[29] N. Zou, X. Zhou, G. Chen, N. M. Andoy, W. Jung, G. Liu, and P. Chen, "Cooperative communication within and between single nanocatalysts," *Nat. Chem.* **10**, 607–614 (2018).

[30] X. Liu, T. Chen, and W. Xu, "Revealing the thermodynamics of individual catalytic steps based on temperature-dependent single-particle nanocatalysis," *Phys. Chem. Chem. Phys.* **21**, 21806–21813 (2019).

[31] W. Li, J. Miao, T. Peng, H. Lv, J.-G. Wang, K. Li, Y. Zhu, and D. Li, "Single-molecular catalysis identifying activation energy of the intermediate product and rate-limiting step in plasmonic photocatalysis," *Nano Lett.* **20**, 2507–2513 (2020).

[32] R. F. Hamans, M. Parente, and A. Baldi, "Super-resolution mapping of a chemical reaction driven by plasmonic near-fields," *Nano Lett.* **21**, 2149–2155 (2021).

[33] R. Ye, M. Zhao, X. Mao, Z. Wang, D. A. Garzón, H. Pu, Z. Zhao, and P. Chen, "Nanoscale cooperative adsorption for materials control," *Nat. Commun.* **12**, 1–9 (2021).

[34] M. B. Roeffaers, G. De Cremer, J. Libeert, R. Ameloot, P. Dedecker, A.-J. Bons, M. Bückins, J. A. Martens, B. F. Sels, D. E. De Vos et al., "Super-resolution reactivity mapping of nanostructured catalyst particles," *Angew. Chem. Int. Ed.* **48**, 9285–9289 (2009).

[35] G. De Cremer, M. B. Roeffaers, E. Bartholomeeusen, K. Lin, P. Dedecker, P. P. Pescarmona, P. A. Jacobs, D. E. De Vos, J. Hofkens, and B. F. Sels, "High-resolution single-turnover mapping reveals intraparticle diffusion limitation in Ti-MCM-41-catalyzed epoxidation," *Angew. Chem. Int. Ed.* **49**, 908–911 (2010).

[36] G. De Cremer, E. Bartholomeeusen, P. P. Pescarmona, K. Lin, D. E. De Vos, J. Hofkens, M. B. Roeffaers, and B. F. Sels, "The influence of diffusion phenomena on catalysis: a study at the single particle level using fluorescence microscopy," *Catal. Today* **157**, 236–242 (2010).

[37] K.-L. Liu, A. V. Kubarev, J. Van Loon, H. Uji-I, D. E. De Vos, J. Hofkens, and M. B. Roeffaers, "Rationalizing inter- and intracrystal heterogeneities in dealuminated acid mordenite zeolites by stimulated Raman scattering microscopy correlated with super-resolution fluorescence microscopy," *ACS Nano* **8**, 12650–12659 (2014).

[38] Z. Ristanović, M. M. Kerssens, A. V. Kubarev, F. C. Hendriks, P. Dedecker, J. Hofkens, M. B. Roeffaers, and B. M. Weckhuysen, "High-resolution single-molecule fluorescence imaging of zeolite aggregates within real-life fluid catalytic cracking particles," *Angew. Chem. Int. Ed.* **127**, 1856–1860 (2015).

[39] A. V. Kubarev, K. P. Janssen, and M. B. Roeffaers, "Noninvasive nanoscopy uncovers the impact of the hierarchical porous structure on the catalytic activity of single dealuminated mordenite crystals," *ChemCatChem* **7**, 3646–3650 (2015).

[40] Z. Ristanović, J. P. Hofmann, G. De Cremer, A. V. Kubarev, M. Rohnke, F. Meirer, J. Hofkens, M. B. Roeffaers, and B. M. Weckhuysen, "Quantitative 3D fluorescence imaging of single catalytic turnovers reveals spatiotemporal gradients in reactivity of zeolite H-ZSM-5 crystals upon steaming," *J. Amer. Chem. Soc.* **137**, 6559–6568 (2015).

[41] Z. Ristanović, A. V. Kubarev, J. Hofkens, M. B. Roeffaers, and B. M. Weckhuysen, "Single molecule nanospectroscopy visualizes proton-transfer processes within a zeolite crystal," *J. Amer. Chem. Soc.* **138**, 13586–13596 (2016).

[42] K. Naito, T. Tachikawa, M. Fujitsuka, and T. Majima, "Single-molecule fluorescence imaging of the remote TiO$_2$ photocatalytic oxidation," *J. Phys. Chem. B* **109**, 23138–23140 (2005).

[43] K. Naito, T. Tachikawa, M. Fujitsuka, and T. Majima, "Real-time single-molecule imaging of the spatial and temporal distribution of reactive oxygen species with fluorescent probes: applications to tio$_2$ photo- catalysts," *J. Phys. Chem. C* **112**, 1048–1059 (2008).

[44] K. Naito, T. Tachikawa, M. Fujitsuka, and T. Majima, "Single-molecule observation of photocatalytic reaction in TiO$_2$ nanotube: importance of molecular transport through porous structures," *J. Amer. Chem. Soc.* **131**, 934–936 (2009).

[45] T. Tachikawa, N. Wang, S. Yamashita, S.-C. Cui, and T. Majima, "Design of a highly sensitive fluorescent probe for interfacial electron transfer on a TiO$_2$ surface," *Angew. Chem. Int. Ed.* **49**, 8593–8597 (2010).

[46] T. Tachikawa, S. Yamashita, and T. Majima, "Evidence for crystal-face-dependent TiO$_2$ photocatalysis from single-molecule imaging and kinetic analysis," *J. Amer. Chem. Soc.* **133**, 7197–7204 (2011).

[47] W. Xu, P. K. Jain, B. J. Beberwyck, and A. P. Alivisatos, "Probing redox photocatalysis of trapped electrons and holes on single Sb-doped titania nanorod surfaces," *J. Amer. Chem. Soc.* **134**, 3946–3949 (2012).

[48] T. Tachikawa, T. Yonezawa, and T. Majima, "Super-resolution mapping of reactive sites on titania-based nanoparticles with water-soluble fluorogenic probes," *ACS Nano* **7**, 263–275 (2013).

[49] J. B. Sambur, T.-Y. Chen, E. Choudhary, G. Chen, E. J. Nissen, E. M. Thomas, N. Zou, and P. Chen, "Sub-particle reaction and photocurrent mapping to optimize catalyst-modified photoanodes," *Nature* **530**, 77–80 (2016).

[50] J. B. Sambur and P. Chen, "Distinguishing direct and indirect photoelectro-catalytic oxidation mechanisms using quantitative single-molecule reaction imaging and photocurrent measurements," *J. Phys. Chem. C* **120**, 20668–20676 (2016).

[51] X. Mao, C. Liu, M. Hesari, N. Zou, and P. Chen, "Super-resolution imaging of non-fluorescent reactions via competition," *Nat. Chem.* **11**, 687–694 (2019).

[52] W.-K. Wang, J.-J. Chen, Z.-Z. Lou, S. Kim, M. Fujitsuka, H.-Q. Yu, and T. Majima, "Single-molecule and -particle probing crystal edge/corner as highly efficient photocatalytic sites on a single TiO$_2$ particle," *Proc. Natl. Acad. Sci. U.S.A.* **116**, 18827–18833 (2019).

[53] W. Xu, H. Shen, Y. J. Kim, X. Zhou, G. Liu, J. Park, and P. Chen, "Single-molecule electrocatalysis by single-walled carbon nanotubes," *Nano Lett.* **9**, 3968–3973 (2009).

[54] M. R. Hoffmann, S. T. Martin, W. Choi, and D. W. Bahnemann, "Environmental applications of semiconductor photocatalysis," *Chem. Rev.* **95**, 69–96 (1995).

[55] B. Huang, W. Wang, M. Bates, and X. Zhuang, "Three-dimensional super-resolution imaging by stochastic optical reconstruction microscopy," *Science* **319**, 810–813 (2008).

[56] P. L. Hansen, J. B. Wagner, S. Helveg, J. R. Rostrup-Nielsen, B. S. Clausen, and H. Topsøe, "Atom-resolved imaging of dynamic shape changes in supported copper nanocrystals," *Science* **295**, 2053–2055 (2002).

[57] B. Zugic, L. Wang, C. Heine, D. N. Zakharov, B. A. Lechner, E. A. Stach, J. Biener, M. Salmeron, R. J. Madix, and C. M. Friend, "Dynamic restructuring drives catalytic activity on nanoporous gold–silver alloy catalysts," *Nat. Mater.* **16**, 558–564 (2017).

Index

Pages in *italics* refer to figures.